7/X/89

THE NORTH AMERICAN GRASSHOPPERS

THE NORTH AMERICAN
Grasshoppers

VOLUME I

ACRIDIDAE

Gomphocerinae and Acridinae

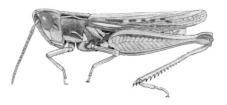

Daniel Otte

Harvard University Press

Cambridge, Massachusetts, and London, England · 1981

Drawings by Daniel Otte
Copyright © 1981 by the President and Fellows of Harvard College
All rights reserved
Printed in the United States of America

Library of Congress Cataloging in Publication Data

Otte, Daniel.
The North American grasshoppers.

Includes bibliographical references and index.
Contents: v. 1. Acrididae: Gomphocerinae and Acridinae.
1. Locusts—North America—Identification. 2. Insects—Identification.
3. Insects—North America—Identification. I. Title.
QL508.A2088   595.7'26   81-6806
ISBN 0-674-62660-5 (v. 1)   AACR2

ACKNOWLEDGMENTS

This work has been made possible by the important collecting, research, and writing done by my predecessors at the Academy of Natural Sciences of Philadelphia: J. A. G. Rehn, M. Hebard, H. Grant, Jr., N. D. Jago, M. Emsley, D. C. Rentz, and H. R. Roberts. Collectors and biologists at other institutions made available an immense amount of information in the form of specimens and published papers. I would also like to acknowledge the outstanding contemporary North American collectors and orthopteran systematists I. J. Cantrall, T. J. Cohn, A. B. Gurney, D. K. McE. Kevan, T. H. Hubbell, and V. R. Vickery. I have greatly benefited from the research on types done by C. S. Carbonell, and I am grateful to D. K. McE. Kevan, V. R. Vickery, I. J. Cantrall, T. J. Cohn, and H. R. Roberts, who read all or major parts of the manuscript and contributed to its improvement.

At the academy my friend and colleague H. R. Roberts has been a constant source of help and encouragement throughout this project. I am also most thankful to Frank Gill and the Academy of Natural Sciences for giving me the chance to practice taxonomy using one of the world's great collections. The help of my assistants has been invaluable. I am especially thankful to my research assistants Donald Azuma, Paulette Francis, Deborah Kotzin, and Katharine Stayman. Elinor Johnson was a most meticulous and helpful typist.

Several institutions lent me specimens, including the British Museum (Natural History); the California Academy of Sciences; The Lyman Entomological Museum, Quebec; the University of Michigan Museum of Zoology; the United States National Museum; Muséum National d'Histoire Naturelle, Paris; Muséum d'Histoire Naturelle, Geneva; and Naturhistoriska Riksmuseum, Stockholm.

Finally, a grant from the National Science Foundation allowed me to initiate this project, and I am thankful to the program directors and reviewers for their support.

Contents

The Subfamily Acridinae   *215*

THE NORTH AMERICAN GRASSHOPPERS

KEY TO NORTH AMERICAN ORTHOPTEROID INSECTS

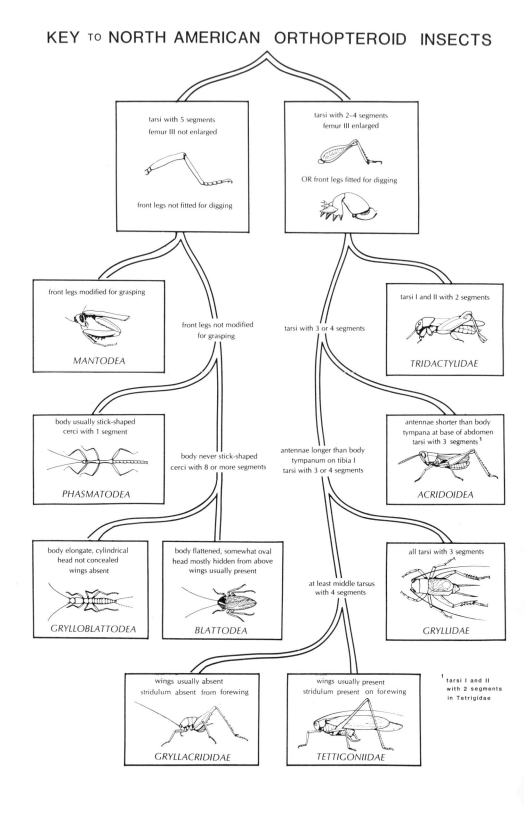

Introduction

The first American grasshoppers were described by Linnaeus in the latter half of the eighteenth century, but the documentation of the fauna did not really begin until a hundred years later. Intensive collecting of North American grasshoppers began in the mid-nineteenth century, and since that time the number of grasshopper species discovered north of Panama has grown to more than a thousand. More precise estimates are not yet possible because the taxonomy of the rich Mexican and Central American faunas is still in a chaotic state.

A count of names does not reflect the actual number of species, because many species were described and named several times. Most species were described between 1860 and 1960 by five men: Samuel Scudder, Lawrence Bruner, Albert Morse, Morgan Hebard, and James Rehn. The history of this activity is reflected in the number of gomphocerine species described during the last two hundred years: 1750–1799, 3 species; 1800–1849, 8 species; 1850–1878, 83 species; 1880–1910, 137 species; 1910–1939, 33 species; 1940 to 1980, 35 species. Although the grasshoppers of the United States have become well known by entomological standards, no comprehensive treatment of the entire fauna exists. In the early 1950s James A. G. Rehn and Harold J. Grant, Jr. began a comprehensive monograph on United States Orthoptera, starting with the Acridoidea. The first and only volume, covering the Tetrigidae, Eumastacidae, Tanaoceridae, and Romaleinae, appeared in 1961. Following Rehn's death in 1965, Grant resumed the work, but the project lapsed after Grant's death in 1967. The present work is not a continuation of their efforts but is rather a fresh start, using a somewhat different approach.

The principal purpose of this work, which will comprise three volumes, is to aid in identifying all described grasshopper species north of the Gulf of Panama, as well as those in the West Indies.

This first volume treats the species of Gomphocerinae and Acridinae, the so-called slant-faced grasshoppers, known from this region. Volume 2 will treat the Oedipodinae, and volume 3, the Melanoplinae, Romaleinae, and other smaller groups.

Grasshoppers are temperate and tropical in distribution, and many are considered pests. About sixteen, mostly Old World, species are nomadic and periodically form large destructive aggregations. The impact of most of the North American species cannot even be roughly estimated, for they have not been studied, but perennially they are rated among the ten worst insect pests. Along with mammals, grasshoppers are the most important grazing herbivores in the world's temperate grasslands, which from man's standpoint are the most important food-producing areas in the world. The competition of grasshoppers with grazing animals for grasses and with humans for grain crops is direct and intense. Their overall impact on the human economy is difficult to assess, for it varies with the region, with the grasshopper species involved, and with the type of food economy. Highly detrimental immediate effects on grazing lands and food crops in certain areas have been well documented, and at the time of this writing, there are major outbreaks in Africa and Asia Minor, with the likelihood of more to come.

The economic importance of grasshoppers means that they must be included in all studies of grassland and desert-grassland communities. Recently they have also been used in studies of the evolution of specialized and generalized diets, the effects of predation on communities of herbivores, and defensive mechanisms evolved by plants against herbivores. They have been used in research on the origin of reproductive isolating mechanisms and on chromosomal mechanisms in relation to speciation. The ecology of insect migrations, of density dynamics, and of large-scale population fluctuations are much better understood as a result of many years of detailed work on economically important species of locusts. It is anticipated that as grasshopper systematics develops, the group will contribute to the knowledge of biogeographic history simply because it is so easily surveyed.

It is probably fair to say that they are biologically better understood than any equivalent group of insects. Because much of the research is comparative, it is important that the studies be based on firm taxonomic underpinnings. Unfortunately, many studies are impeded by a poor taxonomic framework, and others that might be contemplated are not initiated because of difficulties in determining the species. Efforts to understand faunal patterns, biogeographic history, and community ecology, for example, have often been hindered by the difficulties in identifying species, which required the laborious and expensive preparation of specimens to be sent to the few specialists, who are understandably unwilling to devote much time to routine determination of material for others.

Problems in Identification

Grasshoppers display several kinds of variation, which makes classifying them often difficult, and identification a special problem. The sexes may differ in color, wing length, and size; in some species the sexes are so different that they have been placed in different genera, especially when they were not collected at the same time. Most species of the Central and South American genus *Silvitettix,* for example, display pronounced sexual dimorphism, with males often showing bright yellows, greens, and oranges, and females having somber browns and other cryptic hues. Grasshopper taxonomists have through experience learned which groups are most likely to be confusing in this respect, and they are not fooled as often as the earlier taxonomists who frequently did not collect the specimens themselves or who had very few specimens to work with.

Geographic variation is of course a problem in most groups of organisms, but in a diurnal group such as grasshoppers, in which background matching is important, there is likely to be more color variation from one region or habitat to another than is usual for nocturnal Orthoptera. Even within a population, a number of color patterns may be found. In some of the flightless species, as in the genus *Eritettix,* several major color pattern morphs occur, and within each morph a range of color varieties from grays and browns to greens. In some species even the texture of the exoskeleton may vary with coloration, and ridges on the pronotum may be more clearly defined in specimens with strong color transitions (See *Amphitornus,* for example).

Mating Behavior

Behavioral differences are sometimes helpful in distinguishing among grasshopper species or in suggesting phylogenetic relationship (Faber 1953, Jacobs 1953, Otte 1970). Signals by males which attract females or excite them in courtship are especially well differentiated among species and are likely to reveal the presence of previously unknown species.

Fig. 1. Male of *Chloealtis conspersa* in signaling posture.

Pair formation, the initial meeting of a male and female for mating, is achieved either by chance encounter or by signaling. Among the species that signal over some distance, it is usually the male that attracts the female or at least initiates the reciprocal calling and answering sequence that brings the sexes together (Figs. 1, 2). These signals are of two general kinds. Males may either perform flight displays in which they flash their colored hindwings, with or without accompanying snapping noises (crepitation), or they may produce sound by rubbing their hind legs against the forewings (stridulation; see Fig. 3).

Courtship consists of the interactions between male and female after they have sighted one another and before they mate. Mistakes in pairing are common in grasshoppers, although males generally have visual, acoustic, or chemical means of communicating their sex to other males, and these signals inhibit approaching and courtship

Fig. 2. Male of *Ageneotettix deorum* courting female.

by members of the same sex. Such male-to-male signals are commonly observed when several males have approached the same female or when small aggregations of males and females form by chance or by mutual attraction. Females have sometimes evolved signals that are indistinguishable from those of males, which they employ to inhibit courtship and approaching by males.

The evidence that pairing and courtship signals serve as mechanisms for preventing interspecific matings is mostly indirect. Vari-

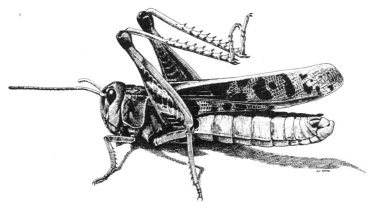

Fig. 3. Male of *Pardalophora haldemanii* (Oedipodinae) signaling with hind femora as he approaches a female.

ous studies have shown that pairing and courtship behavior are dis-
tinctive among closely related species, but that other types of
signaling behavior, such as aggression and repulsion, are relatively
similar. The distinctiveness of the calls of the European Acridinae
facilitated work on taxonomy of that group (Faber 1939) and studies
of behavior are still being used in this way. I believe it likely that the
considerable diversity in courtship signals is caused by selection for
behavioral differences which promote intraspecific recognition,
thereby reducing reproductive interactions between species (Per-
deck 1958).

In one large subfamily, the Melanoplinae, courtship behavior is
quite different. Instead of signaling, the male approaches the female
stealthily, advancing only when the female moves (Fig. 4). Only
after leaping onto the female does the male begin to signal, usually
by shaking its hind legs in a species-characteristic fashion, sometimes continuing until the pair has successfully coupled. Interest-
ingly, in most species that exhibit this type of mating behavior, the
genitalia are highly species-specific.

In all grasshopper species, the male mounts the female to copu-
late and then lowers the end of his abdomen under that of his mate to
attach the genitalia. Insemination takes place by means of a sperm
sac (spermatophore) produced by the male after he attaches to the
female. In some species the shape of the spermatophore is determined by the opening of the male's aedeagus and by the lumen of the
female receptaculum into which the gel-like material is pressed.
After copulation the spermatophore breaks; the female retains the
tube, and the male retains the sac. Until the spermatophore tube is

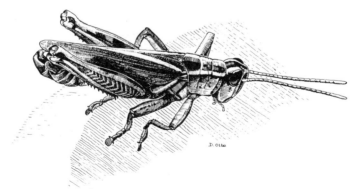

Fig. 4. Male of *Melanoplus bivittatus* (Melanoplinae) preparing
to pounce upon an unsuspecting female.

resorbed by her receptaculum, the female is unreceptive to further copulations.

Some entomologists think, I believe correctly, that genitalic differences can be effective reproductive isolating mechanisms in insects. Others argue that behavioral or other premating differences

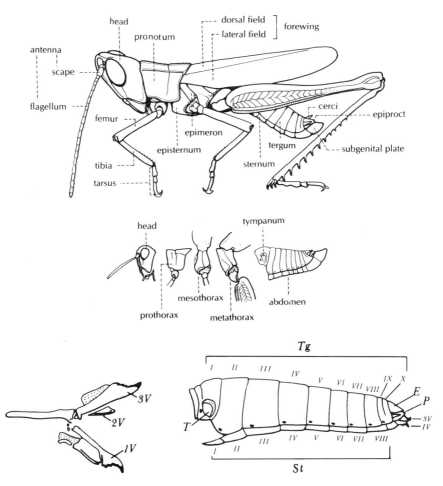

Fig. 5. *Top,* major parts of a grasshopper. *Middle,* body divisions of a grasshopper. *Bottom left,* ovipositor valves: *1V,* ventral valve; *2V,* inner valve; *3V,* dorsal valves. *Bottom right,* female abdomen: *Tg,* tergum with tergites numbered I to X; *St,* sternum with sternites numbered I to VIII; *T,* tympanum; *E,* epiproct; *P,* paraproct.

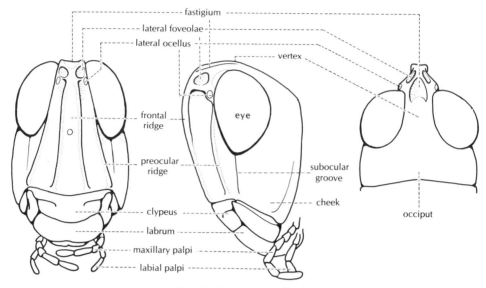

Fig. 6. Grasshopper head.

function as isolating mechanisms and that genitalic differences may be incidental byproducts of selection for increased efficiency in coupling and may not have been selected for reproductive isolation. The latter opinion is based on similarities in genitalia among closely related species and on the fact that even species with different genitalia can cross-mate (Mayr, 1963; Alexander and Otte, 1967).

Grasshopper studies suggest that both theories may be correct; the theory that applies to a particular species depends on that species' mode of signaling. It is interesting that genitalic differences are

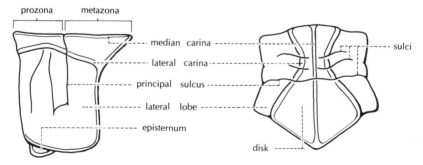

Fig. 7. Grasshopper pronotum.

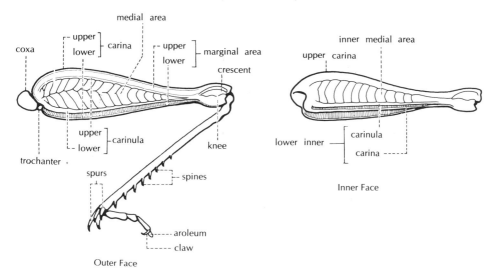

Fig. 8. Grasshopper hind leg.

most pronounced among the Melanoplinae, the group in which males make species-specific signals only after mounting. In this group perhaps genitalic differences facilitate recognition and are directly favored by natural selection to reduce matings with the wrong species. Whether the genitalia serve as true lock and key mechanisms is still problematical, but Cohn and Cantrall (1974), discussing the melanopline genus *Barytettix,* showed that in some cases the genitalia are so different as to make insemination between certain species pairs unlikely or impossible.

Sound Production

The pertinent literature on sound production and hearing in grasshoppers has been reviewed by Kevan (1954), Uvarov (1966), and Otte (1970). Some grasshoppers possess tympana but evidently do not communicate acoustically. Others have well-developed sound-producing mechanisms and do produce acoustical signals but lack a tympanum. Such combinations suggest that each of these characteristics may be gained or lost independently of the other. The complexity of the sound-receiving organ (tympanum) and the similarity of the tympana in various groups suggest that the organ was acquired once and subsequently lost by a number of groups. But the variety of sound-producing, or stridulatory, mechanisms found in

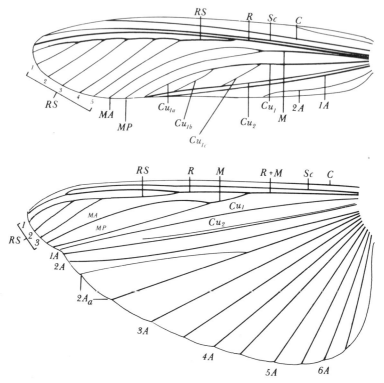

Fig. 9. Major veins of grasshopper wings. C, costa; Sc, sub-costa; R, radius; RS, radical sector (branches numbered); M, media; MA, anterior media; MP, posterior media; Cu_1, anterior cubitus (branches lettered); Cu_2, posterior cubitus; A, anal veins.

grasshoppers, even within a single species, suggests that such mechanisms evolved independently a number of times.

Species that do not signal audibly may produce sounds inaudible to the human ear, or the sounds made by walking, jumping, or flying may have communicative significance. In these species possibly the tympana are used to detect approaching predators. Tympana tend to be lacking in wingless families and to be absent or reduced in short-winged or wingless members of groups that normally have well-developed tympana (Uvarov 1966), suggesting an association of some sort between flight and hearing by means of tympana. Such a secondary reduction of wings and tympana has occurred independently in many groups.

Experimental studies (see Uvarov 1966: 191) show that the tympana of some species are most sensitive to frequencies well within the hearing range of humans, but others are more sensitive to frequencies inaudible to the human ear. Most studies (Haskell 1956, 1961; see also Uvarov 1966 and Michelsen 1966 for references and discussions) also suggest that pitch discrimination is less important than discrimination of the temporal patterns of sound.

■ **Key to Families and Subfamilies of North American Grasshoppers**

1. Pronotum extending over all or nearly all of abdomen (Fig. 10A), fore and middle tarsus each with two segments:
                                                        TETRIGIDAE
   Pronotum short, not covering abdomen (Fig. 10B). Fore and middle tarsus each with three segments: **2**

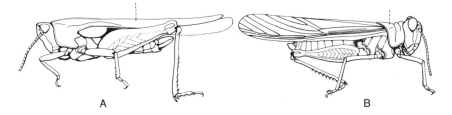

A   B

Fig. 10

2. Antennae shorter than front femora (Fig. 11A, C):   **3**
   Antennae longer than front femora (Fig. 11B):   **4**
3. Body extremely elongated, sticklike. Prothorax tubelike, longer than front femora (Fig. 11C, D). Head conical, with eyes far anterior to mouthparts; front of head anterior to eyes sometimes extremely elongated into rod or cone shape (Central America to southern South America):   **PROSCOPIIDAE**
   Body not extremely elongated or sticklike. Prothorax shorter than front femora (Fig. 11A). Head not conical. Eyes not far anterior of mouthparts: **EUMASTACIDAE**
4. Antennae very long, in males longer than body and in females at least as long as hind femora (Fig. 12A). Males with specialized stridulatory organ on sides of third abdominal segment (Fig. 12B). Frontal ridge forming single carina below ocellus (Fig. 12C). (Rare, known only from extreme southwestern United States and Baja California):   **TANAOCERIDAE**

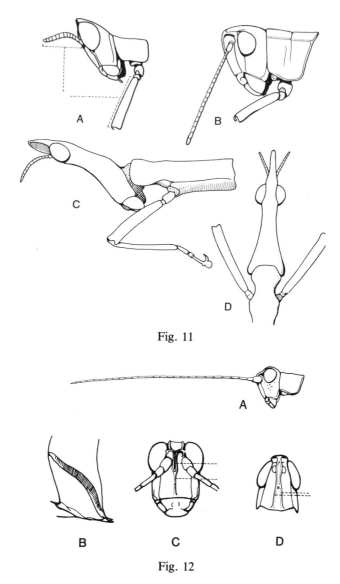

Fig. 11

Fig. 12

Antennae never longer than body. Males without stridulatory organ on side of third abdominal segment. Frontal ridge either broad or with two carinae below ocellus (Fig. 12D):   **5**
5. Male with stridulatory file on side of second abdominal segment. Both sexes entirely wingless. Abdominal tympanum absent. Face incurved in lateral profile. Prosternal spine present.

Head conical. Integument rugose. (Known only from Mexico):   **XYRONOTIDAE**
Males without abdominal stridulatory file and without above combination of characters: **6**
6. Front of head (top view) with median suture dividing fastigium symmetrically (Fig. 13A):   **PYRGOMORPHIDAE**
    Front of head without median suture (Fig. 13B):   **7**

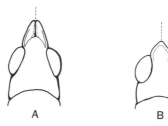

Fig. 13

7. Last outer immovable spine (not spur) located at apex of hind tibiae (Fig. 14A):   **ROMALEIDAE**
    Last outer immovable spine of hind tibiae located some distance from apex (Fig. 14B) (**ACRIDIDAE**):   **8**

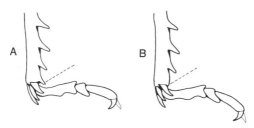

Fig. 14

8. Prosternum between front legs armed with tubercle or flattened process (Fig. 15), and inner face of hind femora always lacking stridulatory file. (Some Gomphocerinae, such as *Paropomala*, Fig. 23A, possess both a protuberance and a stridulatory file. Some Romaleidae possess the protuberance but also have outer apical tibial spine; see 7 above):

Spine-breasted **ACRIDIDAE**
(Ommatolampinae, Rhytidochrotinae,
Leptisminae, Copiocerinae, Proctolabinae,
Melanoplinae, and Cyrtacanthacridinae)

Prosternum between front legs without protuberant process or
spine: other **ACRIDIDAE**
(Gomphocerinae, Acridinae, Oedipodinae)

Fig. 15

● **Identification of North American Acrididae
   Without a Prosternal Spine**

Gomphocerinae (tooth-legged grasshoppers)
   1. Inner face of male hind femora with row of stridulatory pegs
      (Fig. 16A). (Some species of *Orphulella*, the only species of
      Melanotettix, all species of *Stethophyma*, and one species of
      Achurum have evidently lost the pegs secondarily.)

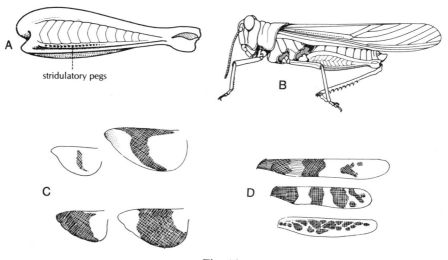

Fig. 16

2. Forewings rounded at apex (see Fig. 21C).
3. Male hindwings without enlarged cells near leading edge (except *Phaneroturis, Orphulella orizabae,* and *Orphulella tolteca*—but these species all have stridulatory pegs).
4. Forewings without raised intercalary vein as in Fig. 16B (dotted line).
5. Hindwings usually transparent (*Acrolophitus* species have hindwings black, or banded yellow and gray; *Syrbula montezuma* has black hindwings; *Bootettix joerni* has pink or reddish hindwings).
6. Forewings without prominent crossbands (not like Fig. 16D), although sometimes spotted.
7. Face (in side view) strongly slanted to vertical.
8. Antennae filiform or ensiform.

Acridinae (slant-faced grasshoppers, American species)
1. Inner face of male hind femora without stridulatory pegs.
2. Forewings obliquely truncated at apex, but truncation more pronounced in males (Fig. 21B) (in *Orphula* females, this feature is sometimes barely discernible).
3. Male hindwings always with enlarged cells near leading edge (Fig. 76).
4. Forewings without raised intercalary vein as in Fig. 16B (dotted line).
5. Hindwings usually transparent or only faintly tinged with color and without dark or smoky band.
6. Forewings without distinct crossbands.
7. Face moderately to strongly slanted.
8. Antennae slightly to strongly ensiform.

Oedipodinae (banded-wing grasshoppers)
1. Inner face of male hind femora without stridulatory pegs.
2. Forewings rounded at apex.
3. Male hindwings always without enlarged cells near leading edge.
4. Forewings usually with raised intercalary vein, as in Fig. 16B (dotted line).
5. Hindwings usually banded or brightly marked with yellows, oranges, reds, or blues and often with black or smoky crossband (Fig. 16C) (clear in some members of *Encoptolophus, Trachyrhachys, Chortophaga,* and *Dissosteira*).
6. Forewings of many species with striking crossbands or large dark patches (Fig. 19D).
7. Face usually more vertical than slanted (except *Psinidia* and *Machaerocera*).

8. Antennae almost always filiform (ensiform in *Psinidia* and *Machaerocera*).

The Gomphocerinae and Acridinae

The subfamilies Acridinae and Gomphocerinae are presently separable only on the basis of the stridulatory apparatus. Species that possess a row of stridulatory pegs on the inner face of the hind femur are assigned to the Gomphocerinae; those that lack pegs are assigned to the Acridinae. When the pegs are secondarily lost, as has happened in several gomphocerine lineages, a species or group may be wrongly assigned to the Acridinae, unless it can be strongly associated with the Gomphocerinae through shared characters.

The members of the Gomphocerinae inhabit a wide range of habitats, from boreal tundra and swamps to extreme deserts and tropical rain forests. Many of the species communicate by means of sound, especially those living in grasslands and open woodlands. Others evidently have secondarily lost their acoustical signals and communicate mainly with conspicuous movements. Species living on the desert floor (*Cibolacris* and *Xeracris*) or open patches of ground in prairies and desert grasslands (*Aulocara*, *Heliaula*, and *Cordillacris*) communicate principally by visual signals, although retaining their stridulatory file and some acoustical signals (Otte 1970).

As a rule, the members of these subfamilies feed on grasses. Most are not specialized feeders, but a few (for example, *Opeia*) feed principally or exclusively on the dominant grass species in their habitat (Mulkern et al 1969; Otte and Joern 1977; Joern 1979). A few desert species (*Cibolacris*, *Ligurotettix*, and *Xeracris*) feed mainly on dicotyledenous shrubs and bushes, and one group (*Bootettix*) feeds exclusively on creosote bushes (*Larrea divaricata*). Most Gomphocerinae reside on vegetation, usually grasses, but some have secondarily become ground-inhabiting and have assumed morphological features resembling those of the mainly ground-living Oedipodinae.

Problems in Classification

There is not yet agreement on how to classify the genera in this volume. Experts tend to cluster most of the genera under one subfamily, but the classification still represents a point of dispute. The one adopted here is a compromise among several previous classifications (see Appendix I).

The apparent ease with which genera lose their pegs and the fact that this is the only reliable character to separate the two subfamilies make it probable that the subfamily Acridinae is polyphyletic. The members of the acridine tribe Hyalopterygini may also be gomphocerines that lost their pegs in an ancestral group. It seems likely that the Acridinae presently include true acridines (lineages that never possessed stridulatory pegs) as well as a number of gomphocerines that have lost their pegs secondarily. If this is so, the group is artificial, in that some of its members are quite unrelated to one another.

The southern subspecies of *Achurum carinatum* from Florida lacks pegs in both sexes, while the northern subspecies retains them. The pegs appear to have been lost independently in at least three different lineages of the tribe Orphulellini, and the only species in the genus *Melanotettix* from Mexico also lacks stridulatory pegs. I place *Stethophyma,* which may have lost its stridulatory pegs secondarily, under the Gomphocerinae, for in other respects (habitat preference, behavior, general morphology) these species seem very similar to bona fide gomphocerines.

Fig. 17 illustrates the geographic pattern of species density of the Gomphocerinae and Acridinae in North America. If the diversity map were superimposed on a vegetation map, it could be seen that the rather sudden transitions to fewer species away from the central part of the continent coincide with the margins of the major forest biomes. Parts of northern Mexico probably have more species than the map indicates, but the region has been less thoroughly collected than adjoining portions of the United States. Central America is also probably underrepresented, and more collecting will doubtless reveal a greater density of species there.

In this volume the genera and species are arranged according to their presumed phylogenetic relationship whenever practical, but distortions and artificialities are introduced when a two- or three-dimensional network is forced into a linear sequence. The arrangement suffers further from the fact that phylogenetic relationships among taxa have yet to be studied. Most of the classification agrees with that developed by grasshopper taxonomists over the last seventy or so years, but grasshopper taxonomists will readily detect (and may oppose) certain departures from the traditional scheme. The arrangement adopted here is based partly on tradition, partly on a consideration of additional evidence, and partly on intuition. Although the field may not yet be close to consensus, I believe the present system will work for the time being.

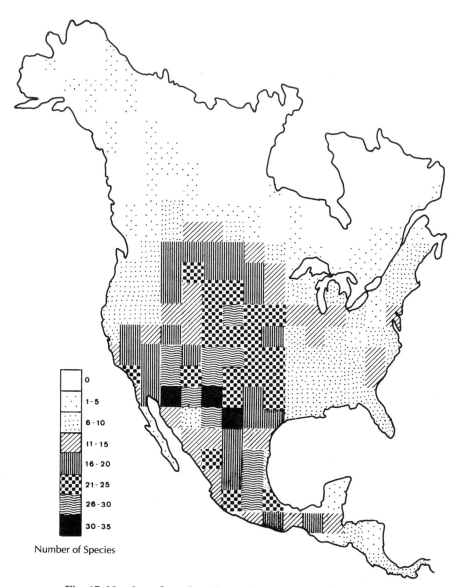

Number of Species

Fig. 17. Number of species of gomphocerine and acridine grasshoppers in North America. Parts of Mexico and Central
America may be underrepresented because of less intensive collecting.

Method of Description

Under the Recognition sections for genera and species, I have listed a number of characteristics by which a taxon may be recognized. I have not listed the characters in a set sequence but roughly in order of importance for identification. Although this is the quickest way to proceed, it does not permit a detailed comparison of morphological features. For some groups, point-for-point comparisons or tables are given. Distribution maps are provided for most species; each dot on the maps represents a collecting locality.

■ Key to Genera of the Gomphocerinae and Acridinae

1. Lateral foveolae of vertex invisible from above, or foveolar area forms right or acute angle with plane of fastigium so it is invisible from above (Fig. 18A, B): **2**

   Lateral foveolae visible from above (Fig. 18C) or, if foveolae are obsolete, foveolar area forms obtuse angle with plane of fastigium so it is visible from above (Fig. 18D): **31**

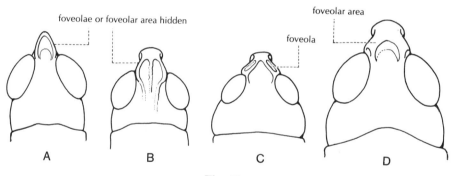

Fig. 18

2. Antennae weakly to strongly ensiform and distinctly wider at base of flagellum than in middle (Fig. 19A, B): **3**
   Antennae filiform or club-shaped (Fig. 19C, D, E); basal segments of flagellum may be slightly flattened but not noticeably wider than distal segments: **19**
3. Lateral margins of pronotal disk nearly parallel and straight Fig. 20A, B) (*Chiapacris* and *Syrbula* males sometimes marginally parallel): **4**
   Lateral margins of pronotal disk constricted in center or diverging on metazona (Fig. 20D, E, F): **12**

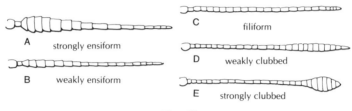

A   strongly ensiform

B   weakly ensiform

C   filiform

D   weakly clubbed

E   strongly clubbed

Fig. 19

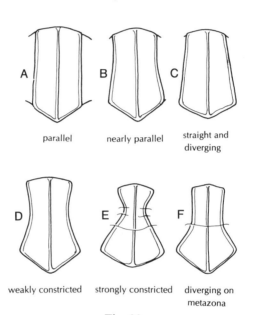

parallel   nearly parallel   straight and
                                       diverging

weakly constricted   strongly constricted   diverging on
                                            metazona

Fig. 20

4. Side of body with strong postocular band. Forewings and hind
   femora extend beyond end of abdomen. Hind tibiae usually
   reddish. Greatest body length at least 25 mm in males and at
   least 32 mm in females: **Mermiria**
   Not fitting above description   **5**
5. Hind femora of male without stridulatory pegs. Ends of forewings obliquely truncated and coming to point at apex (Fig.
   21B):   **Metaleptea**
   Hind femora of male with stridulatory pegs (Fig. 21A). Forewings rounded at apex (Fig. 21C):   **6**
6. Dorsal length of head distinctly longer than front femur: **7**
   Dorsal length of head shorter than or nearly equal to length of
   front femur: **8**

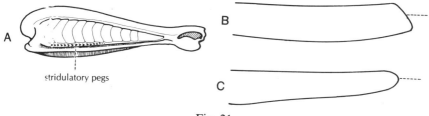

stridulatory pegs

Fig. 21

7. Venter of body with a protuberance between front legs (Fig. 22A). Pronotal disk without lateral carinae. Front of face convex in side view (Fig. 22E, F, G): **Paropomala**
Venter of body without protuberance. Pronotal disk with distinct lateral carinae. Front of face concave in side view (Fig. 22B, C, D): **Achurum**

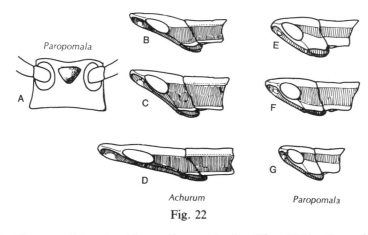

Paropomala

Achurum   Paropomala

Fig. 22

8. Vertex of head with median carinula (Fig. 23A). Forewings usually do not reach beyond apex of abdomen: **9**
Vertex without median carinula (Fig. 23B), although sometimes with very small carinula at front of fastigium (Fig. 23C): **11**
9. End of abdomen extends beyond ends of hind femora (Fig. 24A): **Pseudopomala**
Hind femora extend beyond end of abdomen (Fig. 24B): **10**
10. Lateral pronotal carinae cut by two sulci (Caribbean region and South America): **Leurohippus**
Lateral pronotal carinae cut by one sulcus (Mexico and United States): **Opeia**
11. Forewings extend at least to end of abdomen in both sexes.

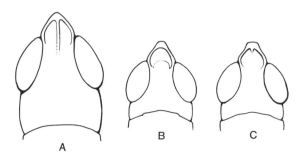

Fig. 23

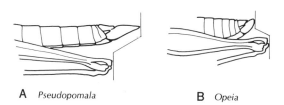

A   *Pseudopomala*   B   *Opeia*

Fig. 24

Opeia obscura

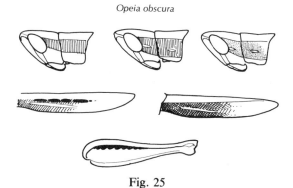

Fig. 25

Pronotal disk with well-developed lateral carinae and without accessory carinulae (Fig. 26C, F):   **Chiapacris**
Forewings do not reach end of abdomen. Pronotal disk often
without lateral carinae and usually with closely spaced parallel accessory carinulae (Figs. 26A, B, D, E):   **Silvitettix**
12. Front and middle femora of male enlarged (Fig. 27A). Pronotal
disk with posterior FDI only. Head and pronotum without
accessory carinulae. Fastigium without prominent median

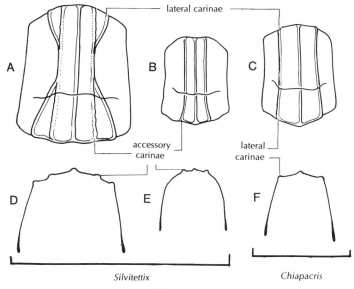

Fig. 26

carinula. Hind femora not banded. Hind tibiae never reddish.
Lateral carinae constricted in central part: **Orphulella**
Front and middle femora not enlarged and otherwise not fitting
above description: **13**

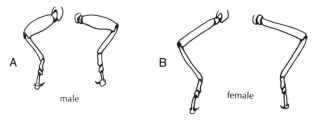

Fig. 27

13. Males without stridulatory pegs on hind femora. Apex of forewings, especially in male, obliquely truncated (Fig. 28A).
Hindwings of male with enlarged cells in anterior lobe (Fig.
28C). Head and pronotal disk without accessory carinulae.
Fastigium without median carinula. Forewing extends beyond end of abdomen in both sexes and is without dark spots
or pale streaks. Lateral pronotal carinae diverge on metazona: **Orphula**

Males with stridulatory pegs. Apex of forewings rounded (Fig. 28B). Hindwings without enlarged cells. Otherwise not fitting above description: **14**

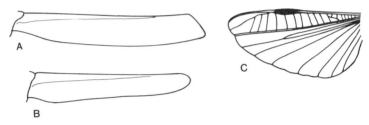

Fig. 28

14. Hind tibiae orange in distal two-thirds and banded with black and white in proximal third. Antennae strongly ensiform and more than one and a half times as long as head plus pronotum. Lateral pronotal carinae cut by three sulci (deserts of southwestern United States and northwestern Mexico):
                                                               Acantherus
    Not fitting above description: **15**
15. Top of head with accessory carinae. Disk of pronotum usually with pale or white lateral carinae and with FDI consisting of continuous bands or two sets of triangles (Fig. 29B, C):
                                                               Eritettix
    Not fitting all of above description: **16**

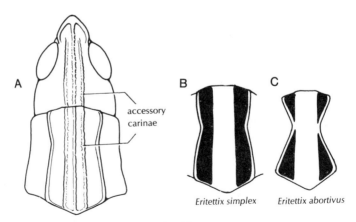

Eritettix simplex   Eritettix abortivus

Fig. 29

16. Forewings short, not reaching end of abdomen; in females often
   shorter than head plus pronotum: **Silvitettix**
   Forewings extending to or beyond end of abdomen: **17**
17. Small gray and white species with clearly defined gray-brown
   postocular band extending to back of pronotum (Fig. 30A,
   B). Antennae pale and moderately ensiform in both sexes.
   Hind femora usually dark in upper half of medial area:
                                               Cordillacris
   Not fitting above description: **18**

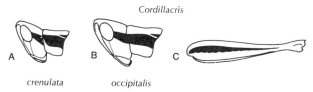

Cordillacris

A   B   C

crenulata   *occipitalis*

Fig. 30

18. Fastigium without median carina. Postocular band covering
   one-half to two-thirds of side of body from back of eye to
   apex of forewing (Figs. 31B, C). Disk of pronotum with, at
   most, tiny posterior FDI: **Chiapacris**
   Fastigium with median carina (Fig. 31A). Side of body without
   postocular band. Disk of pronotum with only posterior or
   with both posterior and anterior FDI: **Syrbula**

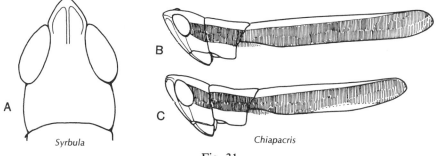

A   B   C

Syrbula   *Chiapacris*

Fig. 31

19. Forewings in both sexes usually shorter than head plus pronotum, usually somewhat pointed apically. Each posterior margin of pronotal disk concave. Stout-bodied, short-winged
   species from central California: **Esselenia**
   Not fitting above description: **20**
20. Body color olive green, with black spots and pearly or irride-

scent white spots on side of body. Pronotal disk mostly reddish on metazona. Pronotal disk strongly diverging on metazona. Lateral pronotal carinae absent. Desert species
inhabiting creosote bushes: **Bootettix**
Not fitting above description: **21**

21. Front of head pointed (Fig. 32A, B). Body hairy. Body color
usually greenish. Side of head with green and pale stripes descending from eye. Hind femora banded with green and
ivory. Antennae long and dark. Hindwings usually marked
with dark pigmentation: **Acrolophitus**
Not fitting above description: **22**

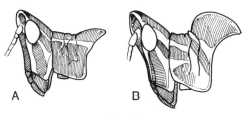

Fig. 32

22. Males with enlarged front and middle femora (Fig. 27A). Hind
tibiae with fewer than sixteen outer spines. Lateral pronotal
carinae well developed. Fastigium without distinct median
ridge. Head and pronotal disk without supplementary
carinae. FDI present or absent at posterior of disk. Hind femora not banded. Lateral field of forewings without pale
streak or row of dark spots: **23**
Males without enlarged front and middle femora. Otherwise not
fitting all of above description: **24**

23. Lateral pronotal carinae slightly to strongly constricted (Fig.
20D, F). Pronotal disk usually with triangular posterior FDI:
Orphulella and **Orphulina** (see page 84)
Lateral pronotal carinae parallel (Fig. 20A). Pronotal disk always without FDI: **Dichromorpha**

24. Forewings short, not reaching end of abdomen, *and* frontal
ridge distinctly grooved next to base of antennae: **25**
Not fitting both of above; either forewings reach end of abdomen or beyond, or frontal ridge is flat or slightly concave
near antennae: **26**

25. Lateral carinae obsolete or indistinct (Fig. 33A, B, D, E) and
often with accessory carinae (Fig. 33B, C, E). Hindwings
without large rectangular cells: **Silvitettix**

Lateral carinae well developed. Accessory carinae lacking (Fig. 33F, H). Hindwings with large rectangular cells (Fig. 33G). Hind femora reddish, unbanded, and with black knees. Posterior margin of pronotum slightly concave at transition from disk to lateral lobes: **Phaneroturis**

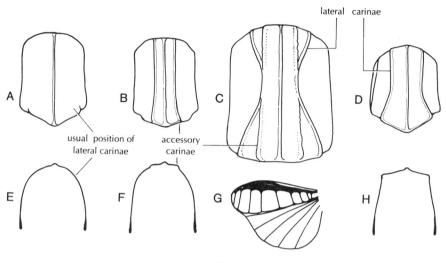

Fig. 33

26. Lateral field of forewings with three to five dark markings (Fig. 34A). Face with vertical ivory band from lower back margin of eye (Fig. 34A). Hind knees of males black. Medial area of hind femora banded: **Phlibostroma**
   Not fitting above description. Face without pale vertical band:

   27

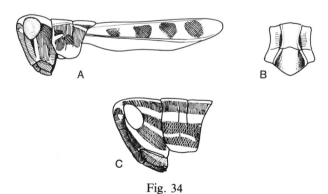

Fig. 34

27. Hind femora strongly banded. Hind margin of pronotum rounded. Lateral lobes and side of head usually marked as in Fig. 34C. Hind tibiae blue. Lateral field of forewings with white streak above base of hind femora. Lateral lobes with horizontal white streak. Disk of pronotum sometimes without lateral carinae and often with two accessory carinulae near median carina: **Amphitornus**
Not fitting above description: **28**

28. Lateral field of forewings with longitudinal white streak above base of hind legs, or with row of large, connecting, dark spots or both. Disk of pronotum with continuous or triangular FDI (Fig. 72E, F, G):   **Syrbula**
Lateral field of forewings more or less unicolorous. Pronotal disk without FDI: **29**

29. Lateral carinae always parallel and cut by two or three sulci. Fastigium in side view rounding gradually onto very broad frontal ridge (Fig. 38C). Hind femora completely unbanded. Forewings in both sexes extend beyond end of abdomen. Top of head often with three closely parallel ridges (Fig. 35A), sometimes with single low ridge (Fig. 35B). Hind tibiae pale brown to dark brown: **Amblytropidia**
Lateral carinae parallel or slightly constricted and cut by one sulcus. Fastigium flat or slightly concave, in side view forming angle with frontal ridge (Fig. 35E). Vertex of head without long median carina (Fig. 35D). Hind femora slightly to strongly banded on external and upper faces. Forewings of males rarely extend beyond abdomen; forewings of females rarely reach end of abdomen. Hind tibiae brownish orange or reddish: **30**

30. Outer face of hind femora (including knees) with three distinct dark bands roughly equal in width. Pronotum of males entirely dark. Hind tibiae yellowish or light brown, becoming blackish at distal end. Forewings of females shorter than head plus pronotum, not overlapping medially (rare, known only from mountains of Idaho):   **Chrysochraon**
Outer face of hind femora may be banded, but not as above. Pronotum of males much lighter on disk than on lateral lobes. Hind tibiae orange or reddish. Forewings of females overlapping medially: **Chloealtis**

31. Pronotal disk with FDI forming triangles (Fig. 36A, B) or elongate bands (Fig. 36C):   **32**
Pronotal disk without FDI: **42**

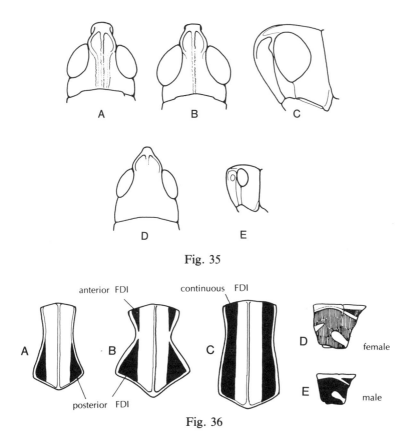

Fig. 35

Fig. 36

32. Hind femora of males entirely black, those of females blackish
    on lower marginal area. Males usually black and with yellow
    green band along dorsum. Females either largely brown or
    greenish. Both sexes with oblique ivory band on lateral
    lobe/metazona (Fig. 36D, E (southwestern Mexico):
                                                    Melanotettix
    Not fitting above description: **33**
33. Body length to end of forewings more than 20 mm in males,
    more than 28 mm in females. Lateral pronotal carinae cut by
    three sulci (Fig. 37B). Lateral foveolae absent (Fig. 37A).
    Frontal ridge convex, rounding gradually onto vertex. Pro-
    notal disk with anterior and posterior FDI (Fig. 36B) (south-
    western United States to South America): **Rhammatocerus**
    Not fitting above description: **34**
34. Antennae strongly clubbed in males; distal segments slightly

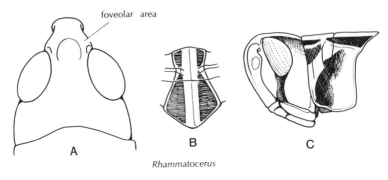

Rhammatocerus

Fig. 37

clubbed and black in females (Fig. 38C). Front tibiae of males greatly enlarged in high-altitude populations (Fig. 38B). Face usually with pale vertical band running from back of antennal socket to anterior articulation of mandible (Fig. 38A) (mountains and northern great plains of United States and Canada):   **Aeropedellus**
Not fitting above description: **35**

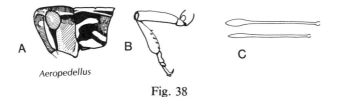

Aeropedellus

Fig. 38

35. Face and sides of body greenish. Abdomen yellow. Forewings uniformly tan to yellowish. Hind femora unbanded. Hind knees black. Frontal ridge grooved. Lateral pronotal carinae well developed and cut by one sulcus. Pronotal disk with only small posterior FDI: **Chorthippus**
    Not fitting above description: **36**
36. Side of head with either black region or black vertical line between eye and front articulation of mandible; in front of this, directly beneath base of antennae, is a vertical ivory band. Lateral lobes mostly dark and with narrow ivory stripe along front margin (Fig. 39A, B):   **Stenobothrus**
    Not fitting above description: **37**
37. Lateral pronotal carinae not forming continuous raised ridge,

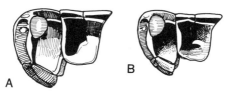

Stenobothrus

Fig. 39

especially indistinct in vicinity of transverse sulci. Lateral
pronotal carinae always cut by three sulci: **38**
Lateral pronotal carinae forming continuous ridge and cut by
one or three sulci: **39**

38. Hind tibiae blue or blue gray. Lateral lobes usually with large
    dark marking, and anterior margin of lateral lobes with pale
    vertical band (Fig. 40A, B):   **Aulocara**
    Hind tibiae orange or red. Lateral lobes without large dark
    marking, usually grayish, and anterior margin not pale:
                                                      Ageneotettix

Aulocara

Fig. 40

39. Lateral lobe with lower ivory band and black or dark upper band
    (Fig. 41B, C, D); ivory band has convexly arching upper
    margin. Top of head usually with accessory carinulae ex-
    tending back from fastigial ridges (Fig. 41A). Antennae
    slightly ensiform. Lateral carinae cut by one sulcus. Top of
    hind femora unbanded or with very indistinct triangular
    marking in central section (southwestern United States and
    western Mexico):   **Horesidotes**
    Lateral lobes not marked as above. Top of head usually without
    accessory carinulae. Antennae filiform. Lateral carinae cut
    by one or three sulci: **40**

40. Lateral field of forewings with three to five large dark spots (Fig.
    34A). Hind tibiae red to orange. Side of head with dark verti-

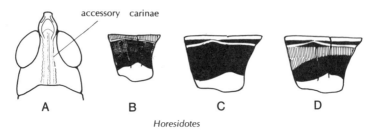

Fig. 41

cal streak from lower margin of eye to anterior articulation of
mandible. Top of hind femora without triangular mark:
                                                            Phlibostroma
Not fitting above description. Top of hind femora with triangular
central band (Fig. 42A):   **41**
41. Antennae pale. Lateral carinae always cut by three sulci. Frontal ridge grooved. Side of thorax not marked as in Fig.
42B–E (genus restricted to California):   **Eupnigodes**
Antennae brownish. Where genus coexists with *Eupnigodes,*
lateral carinae cut by one sulcus. Frontal ridge not grooved.
Side of thorax as in Fig. 42B–E (genus widespread, including
California):   **Psoloessa**

Fig. 42

42. Very small: male body length 11.5 mm to end of hind femora,
female, 17 mm (known only from Cuba):   **Compsacrella**
Body length 20 mm or more (Canada, United States, and Mexico):   **43**
43. Extremely slender species. Subgenital plate in males more than
twice as long as front femora. Face strongly slanting so eyes
lie well in front of mouthparts. Antennae strongly ensiform
and longer than hind femora (Arizona and Sonora):
                                                            Prorocorypha
Not fitting above description: **44**
44. Pronotum without lateral carinae, especially on prozona; disk

merges imperceptibly into lateral lobes (Fig. 43A). Pronotum
diverging strongly on metazona (desert species):   **45**
Lateral carinae clearly indicated, or transition from disk to lateral lobes clearly indicated (Fig. 43B, C):   **48**

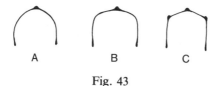

A   B   C

Fig. 43

45. Body gray to dark gray or pale brown on top, blackish on sides.
    Forewings not spotted and with enlarged cells at anterior
    margin (Fig. 44A, B) (living on bushes in Chihuahuan and
    Sonoran deserts):   **Ligurotettix**
    Body color variable but usually pastel, ivory or pale gray blue to
    reddish. Body usually spotted with black, often finely spot-
    ted and sandy-colored (living on ground in southwestern des-
    erts):   **46**

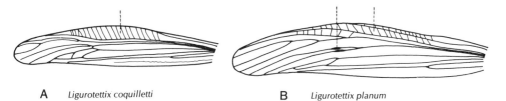

A   *Ligurotettix coquilletti*   B   *Ligurotettix planum*

Fig. 44

46. Hind tibiae orange. Posterior sulcus of pronotum as in Fig. 45A.
    Fastigium of vertex largely convex. Posterior margin of
    pronotal disk usually darkened: **Heliaula**
    Hind tibiae blue, gray, whitish, or yellowish (slightly pinkish in
    one species). Fastigium concave: **47**

47. Hind tibiae usually blue or blue-gray (pinkish in *C. weissmani*).
    Posterior sulcus of pronotal disk as in Fig. 48B, C. Front
    margin of pronotal disk with two bumps. Hind margin of
    pronotal disk angulate (Fig. 45B, C):   **Cibolacris**
    Hind tibiae yellowish or white. Posterior sulcus of pronotal disk
    as in Fig. 49D, E. Front margin of disk without two distinct
    bumps. Hind margin of disk rounded (Fig. 45D, E):

          Xeracris

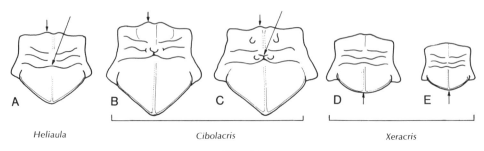

Heliaula   Cibolacris   Xeracris

Fig. 45

48. Top of head broadly rounded in lateral profile (Fig. 46A). Lateral foveolae obsolete or barely indicated. Forewings never extending beyond hind femora and usually not reaching end of abdomen. Hind tibiae usually with black or red coloration, never blue: **Boopedon**
    Not fitting above description: **49**

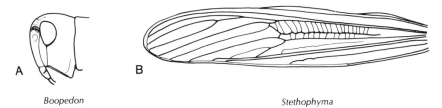

Boopedon   Stethophyma

Fig. 46

49. Hind tibiae blue. Lateral lobes of pronotum with large dark marking in upper anterior half and pale anterior margin (Fig. 40A, B):   **Aulocara**
    Hind tibiae not blue. Lateral lobes not marked as above: **50**
50. Body green or yellow green. Hind femora of males without stridulatory pegs. Males more than 25 mm, females more than 30 mm to end of forewings. Hind knees black. Fastigium of vertex with distinct median carinula. Hind tibiae yellow, sometimes marked with black. Forewings with intercalary vein and connecting cross-veins bearing numerous small teeth or bumps (Fig. 46B):   **Stethophyma**
    Not fitting above description: **51**
51. Hind femora strongly banded on top. Hind knees black. Hind tibiae orange or red. Lateral carinae indistinct and cut by three sulci: **Ageneotettix**

Hind femora unbanded, or with at most a triangular mark on
top. Hind knees not black, hind tibiae yellowish, brownish,
or blackish. Lateral carinae distinct and cut by one sulcus.
Upper half of lateral lobes generally dark, lower half ivory:
Horesidotes

Gomphocerinae

Chrysochraon Genus Group

Two North American genera are tentatively placed in this group. In both, the lateral foveolae are invisible from above, the antennae are filiform, the pronotal disk lacks FDI, the lateral pronotal carinae are parallel or slightly constricted, male forewings rarely extend beyond the abdomen, and female forewings rarely reach the end of the abdomen. The two genera can be distinguished as follows.

• **Identification of Genera**

Chloealtis (widespread over United States and Canada)
   1. Outer face of hind femora unbanded or indistinctly banded
   2. Male pronotum much paler on disk than on lateral lobes
   3. Hind tibiae orange or reddish
   4. Female forewings overlapping dorsally
Chrysochraon (high mountains of Montana and Idaho)
   1. Outer face of hind femora with three distinct dark bands (including knee band) roughly equal in width
   2. Male pronotum entirely dark
   3. Hind tibiae yellowish or pale brown, becoming blackish distally
   4. Female forewings not overlapping dorsally

Genus CHLOEALTIS Harris

DISTRIBUTION. Ranging over most of North America north of Mexico, except the southern and southeastern states.

RELATIONSHIP. The members of the genus appear to represent

the North American radiation of the largely Asian group Chrysochraontes, which includes the Asian genera *Chrysochraon* Fisher, *Podismopsis* Zubovsky, *Euthystira* Fieber, *Mongolotettix* Rehn, and *Pezohippus* Bei-Bienko (Bei-Bienko 1932; Jago 1971). I here follow the scheme of Jago (1969) and treat *Neopodismopsis* and *Napaia* as synonyms of *Chloealtis*.

RECOGNITION. Antennae ensiform to filiform and usually flattened in the first quarter. Lateral foveolae either absent (*C. abdominalis* and *C. conspersa*) or present and visible from above. Frontal costa convex. Fastigium concave, flat, or slightly convex and with a small median carina or raised line. Forewings in both sexes not reaching the end of abdomen (rare exceptions occur in *C. conspersa* and *C. abdominalis*). Side of head and pronotum without white streaks, spots, or bands. Disk of pronotum without FDI, and sides either parallel or constricted. Lateral pronotal carinae parallel or slightly constricted, always well developed and cut by one sulcus. Hind femora without dark bands on upper face, but sometimes indistinctly banded on outer face. Hind tibiae orange or reddish. Forewings usually unicolorous, but sometimes having a finely granular appearance. Hind margin of pronotal disk straight or slightly angulate. Females sometimes similar to *Dichromorpha* females, but in the latter the pronotal carinae are straight and parallel.

REFERENCES. Bei-Bienko 1932, Gurney, Strohecker, and Helfer 1964, Jago 1969.

• Identification of Chloealtis Species

MALES

conspersa (Rocky Mountains to Atlantic coast)
1. Lateral lobes of pronotum entirely black (as in *dianae*)
2. Side of abdomen black, especially near base
3. Hind femora with pale spot on outer face
4. Dorsum of head and pronotum without dark median stripe
5. Lower marginal area of hind femora not reddish
6. Front of mandible not as in *gracilis*

abdominalis (widespread except far west)
1. Lateral lobes of pronotum brownish, darkening dorsally
2. Side of abdomen light brown
3. Hind femora without pale spot on outer face
4. Dorsum of head and pronotum without dark median stripe
5. Lower marginal area of hind femora reddish
6. Front of mandible not as in *gracilis*

dianae (northern California)
1. Lateral lobes of pronotum mostly black, darker dorsally
2. Side of abdomen black, especially near base
3. Hind femur with pale spot on outer face
4. Dorsum of head and pronotum with median dark stripe
5. Lower marginal area of hind femur not reddish
6. Front of mandible not as in *gracilis*

aspasma (southern Oregon)
1. Lateral lobes of pronotum brown
2. Side of abdomen black near base
3. Hind femora with pale spot on outer face
4. Dorsum of head and pronotum without median dark stripe
5. Lower marginal area of hind femora not reddish
6. Front of mandible not as in *gracilis*

gracilis (southern California)
1. Lateral lobes of pronotum gray-brown, becoming darker dorsally
2. Side of abdomen dark
3. Hind femora without pale spot on outer face
4. Dorsum of head and pronotum without dark median stripe
5. Lower marginal area of hind femora not reddish
6. Face strongly slanted so front articulation of mandible is behind posterior margin of eye

FEMALES: *abdominalis and conspersa*

conspersa
1. Side of abdomen black in basal segments
2. Lower marginal area of hind femora tan or brown

abdominalis
1. Side of abdomen pale brown or brown in basal segments
2. Lower marginal area of hind femora pink to red

Chloealtis conspersa (Harris)   Pl. 1

DISTRIBUTION. Widespread over the northern United States, western Canada, and Quebec.

RECOGNITION. Most similar to *C. abdominalis*. Males: Lateral pronotal lobes entirely black (never entirely black in *C. abdominalis*), and disk pale brown or straw-colored. Side of thorax behind pronotum pale brown (*C. dianae* black in this region). Outer face of hind femora banded with gray and ivory; medial area with an ivory or pale spot in the middle section. Hind knees black. Hind tibiae orange to red. Side of abdomen black at base (never black in *C. abdominalis*).

Females: Body color usually brown to gray-brown, often finely speckled. Upper hind corner of lateral pronotal lobes often black. Inner face of hind femora mostly black in proximal half (not banded in *C. abdominalis*). Side of abdomen black in basal half (never black in *C. abdominalis*). Lateral carinae less constricted than in *C. abdominalis*. Outer face of hind femora indistinctly banded and with a pale spot in the center of the medial area.

HABITAT. Associated with dry upland wooded areas and thickets. According to Morse (1920) it has a "preference for bushy pastures and the edges of woodlands, particularly on dry soil, and wherever old stumps and fragments of decaying wood are accessible for its eggs." Blatchley (1920) reported: "It makes its home in thickets, in the borders to open woods, in grassy plots alongside old rail fences, and oftentimes along the borders of streams in woodland pastures, but it is seldom seen in damp localities."

BEHAVIOR. Rehn (1904) reported:

[Near Pequaming, Michigan] females are very easy to capture when found. They are unable to fly, and their jump lacks entirely the elasticity so characteristic of the males, the latter being about the best jumpers of any of the Orthoptera found in the eastern United States. They have a powerful spring, but it is their quickness which renders them so difficult to capture. When alarmed, they are not content with one spring to a place of apparent safety, but jump about with such speed that they can hardly be followed with the eyes, and never cease their rapid succession of jumps until deep under the nearby vines, twigs or dead grasses, where they cannot be induced to stir, and owing to their color, are perfectly safe. I have never seen a specimen where there was not a thick tangle somewhere near in which to hide. I was only able to catch them by listening until one of the musicians gave his *tsikk-tsikk-tssikk,* which was always lustily delivered.

Females oviposit in decaying wood. Scudder (1874: 372) wrote:

The wood must be firm enough to retain the eggs well in place, and soft enough to absorb much moisture in the spring. Upright pieces of timber are never chosen, but rather short sticks of decaying, charred or pithy wood, which can not easily be broken or blown against the rocks. Holes are frequently made three-quarters of an inch deep, and abandoned because the spot proves unsuitable. In a stick about a foot and a half long, and two or three inches wide, I counted 75 borings, only three or four of which had been used as nests. The number of imperfect to perfect holes must be as 25 to one. When a good piece of wood is discovered, the nests are crowded thickly together; and a stick less than two inches in diameter and five inches in length contained 13 completed nests. The holes are pierced at a slight angle to the perpendicular, away from the insect; they are straight for about a quarter of an inch, then turn abruptly and run horizontally along the grain for about an inch.

The eggs (from 10 to 14 in number) are almost always laid in the horizontal portion of the nest; they are cylindrical, tapering toward the ends, but not at all pointed, and measure from five to five-and-a-half millimeters in length, by one-and-one-eighth in breadth; the ends are equally and regularly rounded.

Isolated males call frequently on warm sunny days. They sing a number of songs at one spot and walk about between series. Each song consists of eight to twenty-one strokes of the femora (Otte 1970).

LIFE CYCLE. Egg-overwintering; adult season from July through September.

REFERENCES. Harris 1841, Scudder 1874, Rehn 1904, E. M. Walker 1909, Blatchley 1920, Morse 1920, Hubbell 1922, Buckell 1922, Cantrall 1943, Brooks 1958, Gurney, Strohecker, and Helfer 1964, Otte 1970.

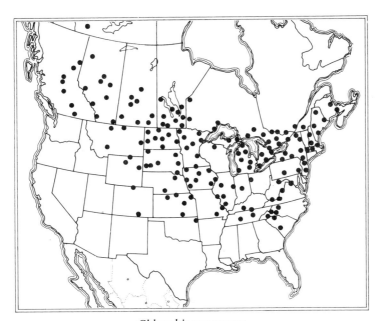

Chloealtis conspersa

Chloealtis abdominalis (**Thomas**)   Pl. 1

DISTRIBUTION. Widespread in forest openings, parklands, and wetter grasslands of Canada, from Ontario to British Columbia and ranging south along the Rocky Mountains to New Mexico.

RECOGNITION. Most similar to *C. conspersa,* but differs in the following ways. Males: Lateral lobes not entirely black, but becoming steadily darker from bottom to top. Pronotal disk more constricted. Hind femora not banded on outer face, reddish or pinkish on lower marginal area. Side of abdomen pale brown at base (black in *C. conspersa*).

Females: Pronotal disk more constricted than in *C. conspersa.* Lower marginal area pink to reddish. Side of abdomen never black. Insides of hind femora brownish and without a black basal area.

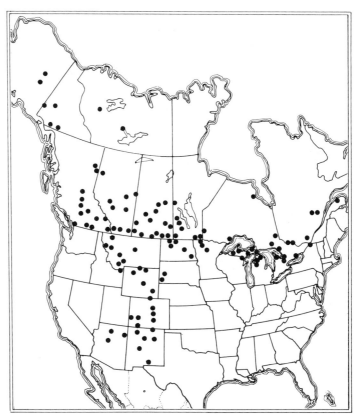

Chloealtis abdominalis

HABITAT. Near Fort Williams, Ontario, Walker (1909) found it "most numerous on the summit of Mt. McKay, where it frequented the small openings in the scrubby woods." In Nebraska it "occurs only in extreme western Nebraska, where it is partial to north hill-slopes and steep canyon walls" (Bruner 1897: 128). At Mammoth Hot Springs, Yellowstone Park, it was found in a small glade at the top of the foothills. Buckell (1922) stated that in British Columbia *C. abdominalis* occurs in the same areas as *C. conspersa* and that females also oviposit in stumps, but it also occurs in open grassy plains (Brooks 1958, Walker 1909). In New Mexico and Arizona it occurs only in moist grasslands at high elevations (Hebard 1935; Ball et al. 1942). I have found *C. conspersa* and *C. abdominalis* intermingling in a grassy glade between the beach and the forest along Lake Superior.

BEHAVIOR. In the field, individual males sing one to three songs at one place and then wander about through the grass before singing again. The songs of solitary males are quite different from those of *C. conspersa* and are composed of about seventeen strokes of both femora. The songs of males singing near other males are both faster and shorter (Otte 1970).

LIFE CYCLE. Adults from July to September.

REFERENCES. Thomas 1873b, Bei-Bienko 1932, Brooks 1958, Gurney, Strohecker, and Helfer 1964, Jago 1969, Otte 1970.

Chloealtis dianae (**Gurney, Strohecker, and Helfer**)   Pl. 1

DISTRIBUTION. Northern coastal ranges of California.

RECOGNITION. Males: Side of body black. Face ivory. Top of body tan or straw-colored and usually with a median dark streak running from front of head to back of pronotum. Forewings mostly pale brown and unspotted, but black below subcosta. Inner face of hind femora black in proximal half; outer face gray brown on medial area, with a small pale spot in the middle; lower marginal area yellow to orange. Hind tibiae orange except for proximal extremity, which is black.

Females: Variable in color; some are colored very much like males, others are straw-colored and lack any black markings on the side of the abdomen. Straw-colored females are difficult to distinguish from females of *C. aspasma,* but the two species evidently do not overlap geographically. Forewings slightly longer than head plus pronotum.

HABITAT. Grassy areas primarily in the eastern part of the northern coast range of California. At Mill Creek, California, the species was found in a small tree-surrounded opening in the canyon bottom, and at Wilbur Springs it occurred only along a short stretch of a grassy, treeless creek bank. East and west of Mendocino Pass and Plaskett Meadows and at Buck Mountain and South Fork Mountain, it inhabited both exposed and tree-shaded areas (Gurney, Strohecker, and Helfer 1964).

LIFE CYCLE. Adults from June through August.

REFERENCES. Gurney, Strohecker, and Helfer 1964, Jago 1969.

Chloealtis aspasma (Rehn and Hebard)   Pl. 1

DISTRIBUTION. Known only from the Siskiyou Mountains, Jackson County, Oregon.

RECOGNITION. Side of body dark brown. Face pale brown. Top of body straw-colored or pale brown. Side of abdomen black at base. Differs from *C. dianae* in having the sides of the body brown and not black, in lacking a median dark stripe on top of the body, and in having small black spots on the upper carinae of the hind femora. Forewings of females about as long as head plus pronotum.

LIFE CYCLE. Adults in the Philadelphia collection were taken in August.

REFERENCES. Rehn and Hebard 1919, Rehn 1928, Jago 1969, Gurney, Strohecker, and Helfer 1964.

Chloealtis gracilis (McNeill)   Pl. 1

DISTRIBUTION. Coastal ranges of southern California.

RECOGNITION. Disk of pronotum with parallel sides. Antennae ensiform. Side of body dark along the top. This darkening is variable, however, and some individuals are scarcely darker on the upper sides. Side of abdomen with broad black markings. Hind femora not banded. Inner face of hind femora not black. Hind knees pale. Hind tibiae brown to orange. Head quite strongly slanted so that an imaginary vertical line running through the front articulation of the mandible passes behind the eye (in other *Chloealtis* it passes through the eye). Male forewings always longer than head plus pronotum; female forewings are either shorter than head plus pronotum or at most slightly longer.

HABITAT. Sometimes found on hillsides with dense bushes and grasses and sometimes in open areas and oak-covered hillsides.

LIFE CYCLE. Adults from May to August.

REFERENCES. McNeill 1897a, Gurney, Strohecker, and Helfer 1964, Jago 1969.

Chloealtis dianae, C. gracilis, and *C. aspasma*

Genus CHRYSOCHRAON Fisher de Waldheim

RELATIONSHIP. *Chrysochraon petraea* is the only known North American member of the genus and was originally the only member of the genus *Barracris* Gurney, Strohecker, and Helfer (1964). But Jago (1971: 251) suggested that *C. petraea* is a relict member of the largely Eurasian group that includes the genera *Podismacris* Bei-Bienko, *Eurasiobia* Bei-Bienko, and *Podismopsis* Zubovsky.

RECOGNITION. See *C. petraea.*

REFERENCES. Jago 1971, Gurney, Strohecker, and Helfer 1964, Dakin 1979.

Chrysochraon petraea (Gurney, Strohecker, and Helfer)   Pl. 1

DISTRIBUTION. Known only from parts of Idaho and Montana.

RECOGNITION. Males: Pronotum entirely dark, lateral lobes black, disk dark gray-brown. Top of head dark brown to black. Forewings uniformly brownish and almost reaching end of abdomen. Hind femora ivory and banded with black on outer, upper, and inner faces; outer face with three oblique black bands (excluding knees); lower marginal area yellowish (orange in females). Hind knees black. Hind tibiae black at base, otherwise brownish and gradually becoming darker until black at the distal end. Lateral pronotal carinae slightly constricted centrally and cut by one sulcus. Body length to end of femora 15–18 mm.

Females: Body color dark gray to black, generally similar to males. Abdomen extending beyond end of hind femora, mottled gray on top, and largely black on the sides. Forewings pointed, not overlapping, shorter than head plus pronotum, and sometimes shorter than the pronotum. Lower marginal area of hind femora orange on both inner and outer faces. Body length to end of abdomen 20–24 mm.

HABITAT. A series of specimens was collected by W. F. Barr above timberline at approximately 9,000 feet. He notes: "They were associated with a large rock slide and were only on and around the rocks. I could not find specimens out in any open area where soil had accumulated and where bits of vegetation were present." (Gurney, Strohecker, and Helfer 1964).

LIFE CYCLE. Specimens were collected July 27, 1961, by W. F. Barr, and on August 19 by Brusven and Gurney.

REFERENCES. Gurney, Strohecker, and Helfer 1964, Jago 1971, Dakin 1979.

Chrysochraon petraea

Chorthippus Genus Group

Only the genus *Chorthippus* from North America is placed in this group, but the genus *Arcyptera* from Asia may also belong. The group is poorly defined and is set up principally because its only North American member does not obviously fit into any other group.

Genus CHORTHIPPUS Fieber

DISTRIBUTION. The genus is represented by perhaps seventy to eighty species, all but one of which are Palaearctic. *Chorthippus curtipennis,* the only Nearctic species, is widespread through the United States and Canada and shows much variation in body size and wing length.

RECOGNITION. See *C. curtipennis.*

REFERENCES. Vickery 1964, 1967.

Chorthippus curtipennis (Harris)   Pl. 1

DISTRIBUTION. This species has an extraordinary distribution in North America, ranging from the tundra of Alaska and Canada to the mountains of Appalachia, New Mexico, and southern California. Not surprisingly, it is morphologically variable across its range, and Vickery (1967) suggested that populations centered in some areas have "diverged to a sufficient degree that they might be given formal subspecies status."

Vickery stated that since *C. curtipennis* in the Rockies is generally found at altitudes of 6,000 feet or more, a series of isolated populations may now be diverging in the west. He also speculated that the species migrated to the Nearctic region during the last interglacial, was pushed to a more southern position during the last glacial, and dispersed northward to its present position as the ice retreated.

Populations from coastal California were given subspecific status (*C. curtipennis californicus* Vickery), because they are smaller, display a greater degree of brachyptery, and many individuals are banded.

RECOGNITION. Commonly greenish on the face and sides of the body, pale brown on the dorsum. Forewings uniformly pale brown to yellowish. Sides of body often greenish. Hind femora pale brown to yellowish and unbanded. Hind knees black, at least on the sides. Hind tibiae yellow to orange, black at base. Abdomen yellowish, banded with black on the sides. Lateral foveolae visible from above, fastigium of vertex slightly convex or flat, sometimes with a small median carina at the anterior end, usually in the form of a raised line. Antennae filiform, becoming black and slightly thicker toward the distal end. Frontal ridge convex. Disk of pronotum constricted on the prozona. Lateral pronotal carinae well developed and cut by one

sulcus. Disk usually with posterior FDI and rarely with small anterior FDI as well. Forewing length variable: in brachypterous individuals east of California, male forewings reach end of the abdomen,
female forewings cover half to three-quarters of the abdomen; in
macropterous males the forewings may be as much as twice as long
as the abdomen and in females one and a half times as long. In California populations (*C. curtipennis californicus*), male forewings may
scarcely extend beyond the middle of the abdomen. In some populations (for example, Fort McMurray, northern Alberta) nearly all individuals are macropterous. At other localities macropterous individuals may predominate in some years and be relatively rare in
others. In 1960 nearly all specimens captured at Sainte Anne de
Bellevue, Quebec, were macropterous, whereas during 1961–1964
macropterous forms were rare.

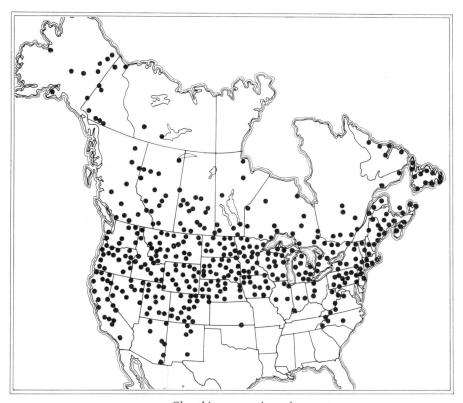

Chorthippus curtipennis

Lateral lobes are variable in color but usually black next to the lateral carinae. Some females are rather distinctly banded on the side of the head and pronotum.

HABITAT. Usually found in medium to tall grasses in moist depressions, swales, edges of marshes and lakes, and reeds. In northern New Mexico and Arizona the species inhabits mountain meadows at 8,000 feet.

BEHAVIOR. Solitary males attract females by stridulating. Responsive females either approach calling males or stridulate in return, causing the males to approach them. The calling song is a series of leg strokes (*zzt-zzt-zzt*) delivered at a rate of about six per second at 80°F (Otte 1970).

LIFE CYCLE. The species overwinters in the egg stage and has an adult season lasting from July to late fall. Kreasky (1960) reports that at high altitudes in the Big Horn Mountains of Wyoming (8,500 ft.), eggs required a three-year development period.

REFERENCES. Blatchley 1920, Morse 1920, Hebard 1935, Ball et al. 1942, Cantrall 1943, Brooks 1958, Vickery 1964, 1967.

Stenobothrus Genus Group

The group includes three North American genera: *Stenobothrus, Aeropedellus,* and *Phlibostroma.* In Asia the genera *Aeropus. Gomphocerus,* and *Gomphocerippus* appear to belong to the group. The three Asian genera and *Aeropedellus* all have moderately to strongly clubbed antennae, and some species have thickened fore tibiae.

The members of this group usually have dark and pale bands descending from the front and lower margins of the eyes and a slanting pale stripe on the lateral lobes.

● **Identification of Genera**

Stenobothrus
  1. Antennae filiform, not clubbed
  2. Fore tibiae not thickened distally
  3. Hind tibiae orange or reddish
  4. Side of forewings spotted or unspotted
Aeropedellus
  1. Antennae club-shaped, especially in males

2. Fore tibiae often thickened distally
3. Hind tibiae brownish or grayish
4. Side of forewings not spotted
Phlibostroma
1. Antennae not club-shaped
2. Fore tibiae not thickened
3. Hind tibiae orange or reddish
4. Side of forewings with three to five large dark spots

Genus STENOBOTHRUS Fisher de Waldheim

DISTRIBUTION. The subgenus ranges from the northwestern United States, through Canada to Alaska. Of the five described species, only three are considered valid, for it is not possible to separate *S. sordida* and *S. alticola* from *S. shastana*. The three valid species are associated with higher altitudes and boreal habitats in the northwestern United States and western Canada.

RELATIONSHIP. *Stenobothrus* is a large, mainly Eurasian genus that includes five subgenera: *Stenobothrus* Fisher, *Omocestus* I. Bolivar, *Myrmeleotettix* I. Bolivar, *Stauroderus* I. Bolivar, and *Bruneria* McNeill (Jago 1971). Only the last subgenus occurs in the Nearctic region.

RECOGNITION. Hind tibiae orange or reddish, at least in distal half. Hind femora banded on upper, outer, and inner faces (inner face only slightly banded in *S. brunneus*). Side of head with a black region or black line between lower side of eye and front of mandible; in front of this is a vertical ivory band directly beneath the base of the filiform antennae. Frontal ridge flat or slightly convex. Back of head often with a pale streak running from the back of the eye to the lateral pronotal carina. Pronotal disk mostly dark, but with a white line along each lateral carina. Pronotal disk slightly constricted in the middle. Lateral carinae cut by one sulcus. Lateral lobes mostly dark, with a narrow vertical ivory stripe along the front margin, and with a small white horizontal streak near the middle of the lobe. The black mark of the lobe is often nearly square but is usually notched at the lower rear corner. Medial area of hind femora usually with three oblique dark bands, but sometimes mostly black. Lower margin of medial area and lower marginal area pale. Hind tarsi orange to red.

REFERENCES. Scudder 1880, McNeill 1897b, Rehn 1906c, Vickery 1969b, Jago 1971.

• **Identification of Stenobothrus Species**

shastanus (western United States)
1. Side of forewings not spotted
2. Forewings not reaching end of abdomen
3. Lateral lobes of pronotum with white marking near center
brunneus (western United States and Canada)
1. Side of forewings spotted
2. Forewings extending to end of abdomen or beyond
3. Lateral lobes of pronotum without white marking near center
yukonensis (Yukon, Canada)
1. Side of forewings spotted
2. Forewings extending to end of abdomen
3. Lateral pronotal lobes without white marking near center

Stenobothrus shastanus **(Scudder)**   Pl. 2

DISTRIBUTION. Northwestern United States.

RECOGNITION. Differing from *S. brunneus* as follows: Forewings in males longer than head plus pronotum, but never reaching end of abdomen. Forewings in females shorter than head plus pronotum and either overlapping slightly or not overlapping medially. Forewings without dark spots. Female abdomen with a strong dark band on the side of each segment, sometimes with a small oblique pale streak above it. Body length to end of femora 15–25 mm in males, 19–26 mm in females.

HABITAT. Grassy meadows and forest openings at elevations from 4,000 to 12,000 feet.

Stenobothrus shastanus

LIFE CYCLE. Adults from mid-July to September.

REFERENCES. Scudder 1880, McNeill 1897a, Scudder 1898, Rehn 1906c, Hebard 1925a, Hebard 1928.

Stenobothrus brunneus (**Thomas**)   Pl. 2

DISTRIBUTION. Northern mountain states and southwestern Canada.

RECOGNITION. Differs from *S. shastanus* as follows: Forewings in both sexes extend to end of abdomen but rarely beyond ends of femora; in egg-laden females, however, the abdomen often extends beyond the wings. Forewings in both sexes with prominent dark spots along the lateral field. Body length to end of femora 16–28 mm in males, 19–28 mm in females.

HABITAT. Grasslands and mountains in northwestern United States and western Canada. In North Dakota Hubbell (1922) found it inhabiting "thin, very dry growth of grasses and other low herbaceous plants." Brooks (1958) reported *Stenobothrus* (*Bruneria*) feeding on several species of *Carex, Bouteloua, Stipa,* and *Agropyron* in Canada.

LIFE CYCLE. Adults are abundant in July, August, and September.

REFERENCES. Thomas 1871a, McNeill 1897a, Vickery 1969b.

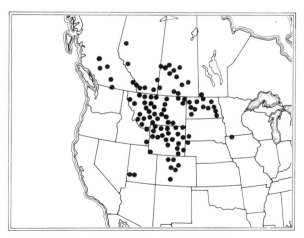

Stenobothrus brunneus

Stenobothrus yukonensis (Vickery)

DISTRIBUTION. Yukon, Canada. To date only three specimens of *S. yukonensis* have been collected, all from Lake Laberge.

RECOGNITION. Very similar to *S. brunneus* but differing in male genitalia (Fig. 47).

HABITAT. This species was found inhabiting the west shore of the lake, in a deep creek valley without trees, and on a grassy slope burned out by the sun.

REFERENCES. Vickery 1969b.

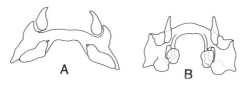

Fig. 47. A, epiphallus of *Stenobothrus yukonensis* from Yukon; B, epiphallus of *S. brunneus* from Chilcotin, British Columbia.

Genus AEROPEDELLUS Hebard

RELATIONSHIP. Ranging from western United States through western Canada to Alaska and Siberia, this genus is apparently related to *Gomphocerus, Gomphocerippus,* and *Aeropus* of Asia. The existence of only two North American species in the genus suggests that the group entered North America relatively recently.

RECOGNITION. Antennae club-shaped (especially in males) with last few segments dark. Front tibiae of males often thickened distally. Lateral pronotal carinae strongly constricted on the prozona and cut by one sulcus.

REFERENCES. Hebard 1935, Bei-Bienko and Mishchenko 1964, Jago 1971.

Aeropedellus clavatus (Thomas)   Pl. 2

DISTRIBUTION. Prairies of Canada and northern United States; high mountains of Colorado, New Mexico, Utah, and Arizona.

RECOGNITION. Antennae club-shaped in both sexes. Front tibiae often thickened in the male. Side of face with a dark streak from the bottom of the eye to the base of the mandible and a vertical white band between this dark line and the preocular ridge. Pronotal disk with triangular posterior FDI. Lateral carinae pale, converging on the prozona, and cut by one sulcus. Lateral lobes with pale diagonal marks in the lower posterior region. Male forewings extend to or beyond end of abdomen. Female forewings usually shorter than head plus pronotum, but longer than pronotum. Body length to end of femora 16–20 mm in males, 18–25 mm in females.

HABITAT. Found on the great plains and ranging to altitudes of 13,000 feet (Alexander and Hilliard 1964). In North Dakota Hubbell (1922) found it common in dry, grassy fields and pastures. He stated: "It was one of the commonest species in the more arid western portion of North Dakota where it occurred abundantly on the dry, grassy uplands and on the grassy slopes and ridges in the Bad Lands." In the prairie provinces of Canada it inhabits dry and somewhat sandy situations south of the forest (Brooks 1958). In Colorado *A. clavatus* feeds principally on grasses and sedges, but alpine populations also feed on the forb *Cirsium hookerianum,* a thistle (Alexander and Hilliard 1964).

LIFE CYCLE. Alexander and Hilliard (1964) studied the life cycle of *A. clavatus* at high altitudes in Colorado. Populations found be-

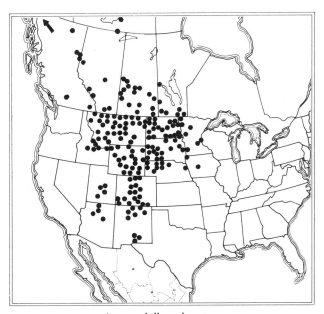

Aeropedellus clavatus

tween 12,800 and 13,100 feet had an abbreviated development, consisting of only four juvenile instars, and hatched while ice and snow were still present. Sexual maturity was reached approximately six weeks after hatching. Laboratory and field evidence indicated that eggs from high-altitude populations passed through two winters before hatching, while those from lower elevations hatched after one winter.

REFERENCES. Thomas 1873a, Hebard 1925b, 1935, Brooks 1958.

Aeropedellus arcticus Hebard

DISTRIBUTION. Alaska and Northwest Territories.

RECOGNITION. Very similar to *A. clavatus*. According to Vickery the antennae are not as distinctly clubbed as in *A. clavatus*, but only slightly enlarged and flattened for some distance; the male forewing length is 9.5–12.0 mm (10.5–14.5 mm in male *A. clavatus*); female forewing length is 8.0–9.6 mm (4.0–7.5 mm in *A. clavatus*). The male epiphallus is also different (see figures in Vickery 1967: 272).

REFERENCES. Hebard 1935, Vickery 1967.

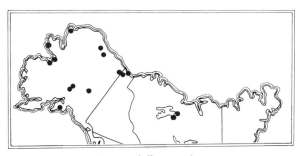

Aeropedellus arcticus

Genus PHLIBOSTROMA Scudder

This monotypic genus is superficially most similar to the genera *Stenobothrus* and *Aeropedellus*. The only species is widespread through the western prairies.

RECOGNITION. See *P. quadrimaculatum*.

Phlibostroma quadrimaculatum (**Thomas**)   Pl. 2

DISTRIBUTION. Short-grass prairies from Canada to Mexico.

RECOGNITION. Lateral foveolae not visible from above. Face with a vertical ivory band descending from lower back margin of eye. Lateral field of forewings with three to five dark, usually semicircular spots, often connected near the top. Forewings variable in length, always longer than head plus pronotum, often not quite reaching end of abdomen, rarely extending beyond hind femora. Hind femora with two dark bands on upper face (excluding knees), and three oblique bands on the medial area. Hind knees of males black, those of females dark on crescent. Hind tibiae orange. Lateral carinae distinct, constricted in the central region, and cut by one sulcus. Posterior margin of pronotal disk usually with triangular FDI on metazona. Lateral lobes with a dark mark of variable shape in the upper two-thirds. Nearly all specimens with two small horizontal

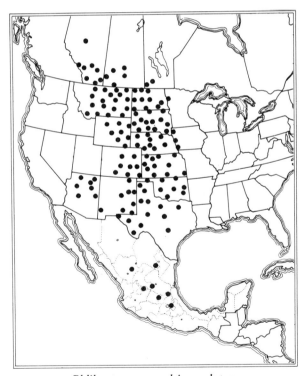

Phlibostroma quadrimaculatum

white streaks on lateral lobes, one connecting bottoms of sulci 2 and 3 and another connecting middle of sulcus 2 to front margin. Body length to end of femora 13–23 mm in males, 21–32 mm in females.

HABITAT. Eroded areas and short-grass prairies. Most commonly associated with clumps of grama grass surrounded by patches of bare ground. The species feeds only on grasses, and at two sites in Nebraska was found to feed predominantly on *Bouteloua gracilis,* although small quantities of six other grass species were eaten (Mulkern et al. 1969). Because it feeds on an important livestock forage grass, the species becomes economically important when it reaches high densities.

BEHAVIOR. Courtship and pair formation are achieved mainly through visual signals and very faint acoustical signals. Males evidently do not have a calling song with which to attract females (Otte 1970).

LIFE CYCLE. Adults from July to September in northern states and into November in northern Mexico.

REFERENCES. Thomas 1871a, Scudder 1875b, Brooks 1958.

Esselenia Genus Group

This group includes only the genus *Esselenia* from North America. Jago (1971: 265) believed that the genus is related to the Tibeto-Chinese genus *Ptygonotus* Tarbinsky.

Genus ESSELENIA Hebard

Esselenia is an aberrant genus with a single species and two subspecies restricted to the coastal ranges of California between Los Angeles and San Francisco.

RECOGNITION. See *E. vanduzeei.*

REFERENCES. Rentz 1966.

Esselenia vanduzeei Hebard   Pl. 2

DISTRIBUTION. Coastal ranges of California. Rentz (1966) reported: ''At present the two subspecies (*E. v. vanduzeei* Hebard and *E. v. violae* Rentz) have not been collected together. Specimens collected in San Benito County, 3 airline miles south of Paicines, have reddish abdomens and reddish hind tibiae, but the distal half of the hind tibiae of males is quite blackish, indicating an intermediate condition.''

RECOGNITION. Stout-bodied and short-winged. Lateral foveolae not visible from above. Forewings usually shorter than head plus pronotum, usually somewhat pointed at the ends, sometimes overlapping medially, but sometimes neither overlapping nor touching. Antennae filiform or at most slightly flattened near base. Left and right rear margins of pronotal disk concave. Lateral carinae strongly constricted at the first sulcus and diverging widely back of this point. Lateral carinae irregular and sometimes lumpy between first and third sulcus. Disk often with large posterior and small anterior FDI. Some individuals almost unicolorous pale tan or gray brown. In individuals with contrasting markings, there are pale lines running along the lateral carinae, and the lateral lobes have a white line from about the middle to the lower back corner. Fastigium very deep and usually with a small median ridge. Frontal ridge strongly grooved. In life *E. v. vanduzeei* has black hind tibiae and no reddish markings on the sides of the abdomen. In *E. v. violae,* the tibiae and sides of the abdomen are bright orange, especially in males. Body length to end of femora 16.5–20.5 mm in males, 20.5–27.0 mm in females.

HABITAT. The species occurs in the Upper Sonoran life zone in areas with reduced rainfall. Specimens often occur in grass among digger pines and oaks. *E. v. vanduzeei* occurs in more hilly areas, while *E. v. violae* is found in more open situations away from dense brush and trees. *E. v. vanduzeei* also has a more southern distribution.

LIFE CYCLE. Rentz (1966) reports that this species is rare in collections because adults emerge early in the spring. Eggs hatch in late winter, nymphs appear in late winter, and adults appear in late March and disappear by early June.

REFERENCES. Hebard 1920, Rentz 1966.

Esselenia vanduzeei subspecies

Melanotettix Genus Group

The group contains one genus and one species, *Melanotettix dibelonius*. This species is in some ways similar to the *Aulocara* genus group, but differs in having lost its stridulatory teeth. The genus is tentatively placed under the Gomphocerinae because of its greater resemblance to gomphocerines than to acridines.

Genus MELANOTETTIX Bruner

RELATIONSHIP. This genus includes but one species, which is restricted to the slopes of southwestern Mexico. Jago (1971: 233) thought the genus represented a brachypterous, flightless isolate of the South American genus *Fenestra*. However, the resemblance does not seem to me close enough to place the two genera in the same group.

RECOGNITION. See *M. dibelonius*.

REFERENCES. Bruner 1904, Jago 1971.

Melanotettix dibelonius Bruner   Pl. 2

DISTRIBUTION. States of Guerrero and Michoacan, Mexico.

RECOGNITION. Males: Body usually black and with a yellow green band on dorsum. Some males chocolate brown. Forewings always shorter than head plus pronotum and sometimes shorter than pronotum. Dorsum of abdomen with a broad brown band; rest of abdomen black. Lateral foveolae small, triangular, and visible from above. Fastigium slightly concave or flat or slightly convex in cross-section and with a slightly raised median ridge. Antennae filiform, black, longer than head plus pronotum. Pronotal disk strongly constricted centrally and with prominent triangular FDI on metazona. Lateral carinae distinct in the anterior third, then disappearing posteriorly. Margin of pronotal disk cut by one sulcus. Lateral lobes dark matte brown or matte black and with a pale oblique stripe running from the center of the lobe to the lower back corner. Hind femora entirely black. Hind tibiae black, but sometimes bluish in central region. Body length to end of femora 24–29 mm.

Females: Tan or pale green on the dorsum. Lateral lobes with an oblique ivory band in the lower back quarter. Upper face of hind femora with two dark marks, a triangular one in the middle and a smaller, less regular shaped one near the anterior end. Lower marginal area of hind femora usually blackish and medial area either brownish or greenish. Hind tibiae blackish or bluish. Body length to end of femora 33–37 mm.

HABITAT. Tall weeds along the roadside and at the edges of old fields (Roberts personal communication).

REFERENCES. Bruner 1904.

Melanotettix dibelonius

Silvitettix Genus Group

Three North American genera (*Silvitettix, Chiapacris,* and *Phaneroturis*) and one South American genus (*Compsacris*) are tentatively placed in one group. All four genera are from forested or wooded areas of the Neotropical region. Males in these genera generally have bright colors and unbanded hind femora.

● **Identification of Genera**

Phaneroturis
1. Hindwings with enlarged cells in anterior lobe (Fig. 48C)
2. Lateral carinae of pronotum well developed and with nearly parallel lateral carinae (Fig. 48A, B)

Silvitettix
1. Hindwings without enlarged cells in anterior lobe
2. Pronotal disk of males usually without lateral carinae; these are replaced by accessory carinae in some species. Females sometimes with both lateral and accessory carinae (Fig. 26)

Chiapacris
1. Hindwings without enlarged cells in anterior lobe
2. Lateral carinae of pronotum well developed and nearly parallel

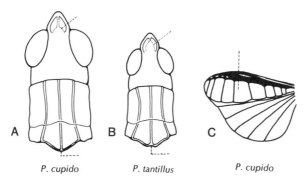

Fig. 48. Comparison of *Phaneroturis* species.

Genus PHANEROTURIS Bruner

DISTRIBUTION. Southern Mexico and Guatemala.

RELATIONSHIP. The two members of this genus appear to be most similar to *Chiapacris* species. Both genera have nearly parallel and clearly defined lateral carinae. Both lack accessory carinae as seen in *Silvitettix*. But *Phaneroturis* males have enlarged cells on the hindwings (Fig. 48C). The forewings are held tentlike over the abdomen, almost like some species of *Chloealtis*.

RECOGNITION. Body length to end of femora 13–20 mm in males, 18–26 mm in females. Forewings short in males, never reaching the end of the abdomen; in females, forewings are shorter than head plus pronotum and not overlapping medially. Hindwings in males with large rectangular cells in the anterior lobe. Hind femora reddish, unbanded, and with black knees. Venter of abdomen yellowish, sides black, dorsum brownish. Fastigium slightly concave or flat in males, flat or convex in females and often with a small median carinula. Disk of pronotum with slightly incurving lateral margins. Lateral carinae distinct and cut by one sulcus. Posterior margin angulate and with small dark triangles, sometimes barely visible in males.

REFERENCES. Otte 1979b.

• Identification of Phaneroturis Species

cupido
1. Male antennae longer than head plus thorax plus wings
2. Male hind tibiae banded black and yellow

3. Female hind knees black
4. Female hind femora mostly reddish brown
5. Narrowest part of hind femora with yellow band
tantillus
1. Antennae shorter than head plus thorax plus wings
2. Male hind tibiae red brown
3. Female hind knees not black
4. Female hind femora greenish, yellow below
5. Narrowest part of hind femora without yellow band

Phaneroturis cupido **Bruner**   Pl. 4

DISTRIBUTION. Southern Mexico and Guatemala.

RECOGNITION. Males: Forewings bright green, with rounded posterior margins, not reaching end of abdomen, and held tentlike over abdomen. Head green, pronotum dark green. Antennae black and longer than head plus pronotum, and middle segments are more than twice as long as wide. Hind femora orange with black knees and with a yellow band on the narrowest part. Hind tibiae mostly blackish but with a yellow band near knees. Side of head with a vertical black line between bottom of eye and anterior articulation of mandible. Body length to end of femora 16–20 mm.

Females: Head, pronotum, and forewings either green or brown. Forewings shorter than head plus pronotum, somewhat pointed, and nonoverlapping. Side of abdomen black. Side of head behind eyes sometimes black. Lateral lobes always black on upper side along lateral carinae. Body length to end of femora 21–26 mm.

HABITAT. Mountain ridges and slopes, often in pine and oak woodlands with grassy or weedy undergrowth (Cantrall field notes).

LIFE CYCLE. Adults in University of Michigan collection were taken in September and February.

REFERENCES. Bruner 1904.

Phaneroturis tantillus **Otte**   Pl. 4

DISTRIBUTION. Southern Mexico and Guatemala.

RECOGNITION. Males: Top of body pale green, occasionally tan. Side of body dark brown to dark gray. Hind femora and hind tibiae reddish, without pale band around narrowest part as in *P. cupido*, and with black knees. Face yellowish, side of head blackish. Fore-

wings dark on sides, green on dorsum, shorter than head plus pronotum. Dorsum of abdomen brownish, sides black, venter yellowish.
Antennae filiform, slightly flattened at base. Body length to end of
femora 13–15 mm.

Females: Variable in color, but usually green. The following
color morphs are known: (a) head, thorax, wings, and hind femora
green; (b) like *a*, but with pink on dorsal surface of wings; (c) like *a*, but
with brown hind femora; (d) head, thorax, and wings brown to dark
brown; (e) like *d*, but dorsum of forewings pink; (f) head and pronotum banded, lateral lobes with two dark and two pale horizontal
bands. Females never with green femora and usually with pale
bands around the narrowest part and on hind tibiae next to knees.
Body length to end of femora 18–22 mm.

HABITAT. High mountain slopes and saddles of mountain ridges.
Inhabits both open meadows and forest undergrowth in pine forests
(Cantrall field notes).

REFERENCES. Otte 1979b.

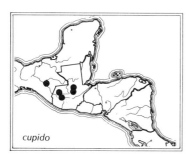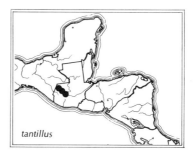

Phaneroturis species

Genus CHIAPACRIS Otte

DISTRIBUTION. Western and southern Mexico.

RECOGNITION. Antennae either flattened in the basal segments or
filiform. Lateral foveolae small or absent, not visible from above
when present. Fastigium concave; transverse arcuate groove in the
anterior half of fastigium. Anterior of vertex sometimes with a very
small median ridge. Top of body pale green or pale brown. Side of
body with a broad black band covering upper one-half to two-thirds
of the lateral lobes and running from the back of the eye to the apex

of the forewings. Forewings extend beyond middle of abdomen. Hind femora without dark crossbands. Knees brown or black. Disk of pronotum with nearly parallel sides, widening slightly on the metazona. Hind margin of disk angulate. Lateral carinae distinct and cut by one sulcus.

REFERENCES. Otte 1979b.

● **Identification of Chiapacris Species**

velox (southern Mexico and Guatemala)
    1. Forewings extending beyond end of abdomen
    2. Postocular dark band covering about half of lateral lobe
    3. Narrowest part of hind femora without yellow band
nayaritus (western Mexico, Nayarit)
    1. Forewings not extending beyond abdomen and about as long as hind femora
    2. Postocular dark band covering about half of lateral lobe of pronotum
    3. Narrowest part of hind femora with yellow band
eximius (central Mexico, Michoacán)
    1. Forewings not reaching end of abdomen, shorter than hind femora
    2. Postocular dark band covering more than 70 percent of lateral lobe
    3. Narrowest part of hind femora without yellow band

Chiapacris velox **Otte**   Pl. 4

DISTRIBUTION. Southern Mexico and Guatemala.

RECOGNITION. Top of body pale green or pale brown. Forewings usually extending beyond hind femora. Antennae moderately ensiform. Hind knees brown or black. Hind femora without a pale ring around the shaft. Hind tibiae brownish, becoming black distally. May be confused with *Orphula azteca,* which differs from *C. velox* in lacking stridulatory pegs in males and in lacking a prominent black band on the side of the body in both sexes. Differs from *C. nayarius* principally in that the wings extend beyond the end of the hind femora. Body length to end of forewings 20–22 mm in males, 24–29 mm in females.

REFERENCES. Otte 1979b.

Chiapacris nayaritus Otte   Pl. 4

DISTRIBUTION. Known only from two males collected in the state of Nayarit—one from San Blas and the other from Campostella.

RECOGNITION. Males: Body color bright green, yellow, and black. Dorsum bright green. Sides of body with a postocular black band running to the apex of the forewings. Antennae black, slightly flattened near the base, and much longer than head plus pronotum. Face, cheeks, and lower third of lateral lobes pale green. Hind femora orange yellow, with black knees and a yellow ring around the narrowest part. Hind tibiae black at base, then mostly yellow, and finally becoming black in fourth quarter. Abdomen bright yellow. Body length to end of femora 22–23 mm in males. Females not known.

REFERENCES. Otte 1979b.

Chiapacris eximius Otte   Pl. 4

DISTRIBUTION. Collected only at Cerro Tancitaro, 2 km north of Apo, Michoacan, Mexico.

RECOGNITION. Males: Dorsum of head, thorax, and abdomen pale green. Side of body largely black. Disk of pronotum with very small posterior FDI and posterior margin slightly angulate. Venter of abdomen yellow. Top three-quarters of lateral lobes black, bottom quarter ivory. Hind femora reddish brown with black knees and without a pale ring on the narrowest part. Hind tibiae black at knee, mostly reddish brown, but becoming dark brown to black in the last quarter. Females not known.

HABITAT. Hubbell described the habitat as follows: "Elevation

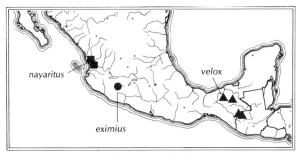

Chiapacris species

2,000–2,100 meters, oak-pine-clad slopes and slope at head of barranca; place where many trees had been cut, some recently, and there was much dry brush and slash in the tall grass.''

LIFE CYCLE. The only existing museum specimens were collected in June 1947.

REFERENCES. Otte and Jago 1979.

Genus SILVITETTIX Bruner

DISTRIBUTION. Widespread through Neotropical forests from Veracruz and Oaxaca to Brazil. The genus presently includes nineteen species—seven from South America and twelve from Central America and southern Mexico. All are associated with forests, forest openings, and forest margins, and many occur in mountainous regions. The ranges of most species are small compared to those of longer-winged species such as *Orphulella*. Such a restriction in range appears to be a direct consequence of their flightlessness. There is no evidence in this tropical group of the occasional flying forms one finds in many predominantly flightless temperate groups. The reason for flightlessness remains problematical, but the answer may lie in the permanence of the habitat or in the fact that the species tend to be restricted to ''islands'' of suitable habitat within forests and so are subject to some of the same pressures promoting winglessness on oceanic islands. With loss of flight and the consequent reduction in vagility, isolated populations are less likely to exchange genetic material over the longer term and thus may speciate more rapidly. Because of the restricted ranges of many species, I believe many more species will be discovered when forest and mountain habitats are more thoroughly explored.

RECOGNITION. Most *Silvitettix* species display pronounced sexual dimorphism, with males often bearing bright yellows, greens, and orange colors, and females having somber browns and other cryptic hues. Differences between the two sexes are so great that early workers placed them in different genera, and even now it is difficult to identify females if they cannot be associated with males. Among other Gomphocerinae, females differ from males mainly in size.

In both sexes of all species the forewings do not reach the end of the abdomen. In most species the lateral carinae of the pronotum in males is replaced by a median pair of accessory carinae flanking the median carina (Figs. 26, 33). In some species males lack all traces of both lateral and accessory carinae, so that the disk rounds gradually onto the lateral lobes. Within species females vary considerably in the development of pronotal carinae; some possess only a pair of strongly bent lateral carinae, and still others, usually those with a pale median band on the dorsum, possess only accessory carinae. The frontal costa in both sexes is moderately to deeply grooved. The lateral foveolar area is always hidden when viewed from above.

■ Key to Silvitettix Species

MALES

1. Side of head mostly one color—brown, green, or black—or with indefinite spots or streaks; sometimes with narrow postocular dark band near top of head: **3**

    Side of head pale below and with sharp transition to broad darker area behind eye: **2**

2. Face sooty black or dark brown in lower part and cream above. Femora all dark green, but hind femora yellow below. Side of body with light cream band occupying lower third of lateral pronotal lobe: **audax**

    Lower part of face pale. Light band on lower half of lateral pronotal lobe. Upper half of lobe brown with upper edge black:

    aphelocoryphus

3. Entire lateral pronotal lobe uniform black or dark brown. Hind femora deep to light purple, red, or pinkish. Face and cheeks black. With or without light yellowish or pale brown dorsum. Forewings varying from pale yellowish or pale brown to purplish black: **communis**

    Lateral pronotal lobe grayish, becoming darker dorsally and with darker band of black, dark brown, or gray at upper margin next to accessory carinae. Otherwise not fitting above description: **4**

4. Length of vertex in front of compound eyes about one and a half times width, seen from above (see Fig. 53E for measurement):   **5**

    Length of vertex in front of compound eyes about the same as width, seen from above: **7**

5. Transverse arcuate groove of vertex crossing fastigium near or at front margin (Fig. 53A, D). Surface of vertex markedly convex or flat. Inner area of hind femora without large black mark: **6**

    Transverse arcuate groove of vertex crossing fastigium at or

TABLE 1. Comparison of North American *Silvitettix* species, males.

Species	Color of hind femur (medial area)	Color of abdomen venter	Position of arcuate groove of fastigium	Shape of rear margin of pronotal disk	Number of stridulatory teeth	Other features
communis	Purple, reddish, or pink	Yellow, orange, or brown	Near middle	Truncated or rounded	About 50	Face and cheeks black
biolleyi	Brown	Orange	Near middle	Rounded or angulate	About 85	
rhachyco-ryphus	Rusty brown or pale green	Yellowish to orange	Front	Nearly truncated	53–58	
whitei	Pale brown	Pale brown	Front	Slightly angulate or rounded	95 (n = 1)	
ricei	Gray above, yellowish below	Yellowish to orange	Near middle	Truncated, centrally concave	About 65	Forewings shorter than head plus pronotum
maculatus	Gray green	Orange	Near middle	Angulate	About 53	Forewings shorter than head plus pronotum
audax	Greenish	Orange	Near front	Slightly angulate, centrally concave	About 75	Top three-quarters of lateral lobes black
aphelocory-phus	Rusty brown	Orange	Front	Slightly angulate	About 65	Top half of lateral lobes darker
salinus	Green or pale brown	Yellow or yellow green	Front	Truncated, centrally concave	?	Narrowest part hind femora orange
chloromerus	Pale green	Yellow or orange	Near middle	Truncated or slightly rounded, centrally concave	About 70	Hind tibiae green and without pale annulus near base
thalassinus	Pale green or yellow	Orange	Near middle	Angulate	100–120	
gorgasi	Brown	Brownish	Near middle	Slightly angulate	63 (n = 1)	Disk of pronotum rough medially

TABLE 2. Comparison of North American *Silvitettix* species, females.

Species	Position of arcuate groove of fastigium	Forewings overlapping on dorsum	Forewings shorter than pronotum	Shape of rear margin of pronotal disk	Front of head with median carinula
communis	Near middle	Usually not	Often	Truncated, rounded, or slightly angulate	Sometimes traces
biolleyi	Nearer middle than front	No	Usually	Angulate	Traces
rhachycoryphus	Front	No	Yes	Truncated and usually concave at center	Traces
whitei	Nearer front than middle	No	Yes	Truncated and usually concave at center	No
ricei	Near front	No	Yes, shorter than head	Truncated, concave at center	No
maculatus	Nearer middle than front	No	Yes	Angulate	Traces
audax	Nearer middle than front	Yes	No	Angulate but center concave	No
aphelocoryphus	Front	Yes	Sometimes	More rounded than angulate, center concave	No
salinus	Front	No	Yes, sometimes shorter than head	Truncated or slightly concave	Yes
chloromerus	Nearer middle than front	No	Yes, shorter than head	Truncated or very slightly rounded	Trace
thalassinus	Nearer middle than front	No, sometimes touching	Yes	Angulate	No

near midpoint. Fastigium decidedly concave. Inner face of hind femora and lower inner carinae with black basal marking, outer area lightly olivaceous, and knees black:

maculatus

6. Vertex convex with distinct median dorsal carinula (Fig. 53A). Hind femora pale green in proximal half, orange in distal half; or brown with sooty streak basally and reddish distally. Knees not black. Hind tibiae blue or gray: **salinus**

Vertex slightly concave and without median dorsal carinula (Fig. 53D). Hind, middle, and fore femora rich green. Hind femora yellow below, knees black. Hind tibiae green, sooty in distal half: **chloromerus**

7. Transverse arcuate groove at or near front of fastigium: **10**
   Transverse arcuate groove near middle of fastigium: **8**

8. Hind femora yellow, yellow green, or pale green: **thalassinus**
   Hind femora brown: **9**

9. Venter of abdomen brownish. Rear margin of pronotal disk angulate. Supplementary carinae indistinct, area between them roughly textured: **gorgasi**
   Venter of abdomen orange. Rear margin of pronotal disk rounded to slightly angulate. Supplementary carinae distinct:
                                                              biolleyi

10. Forewings shorter than a head plus pronotum. Medial area of hind femora dark in upper half: **ricei**
    Forewings longer than head plus pronotum. Medial area of hind femora not dark along upper side: **11**

11. Venter of abdomen pale brown. Stridulatory file of hind femora with over eighty teeth. Frontal ridge strongly constricted above median ocellus (Fig. 51B). Hind femora pale brown:
                                                                whitei
    Venter of abdomen orange. Stridulatory file of hind femora with less than seventy teeth. Frontal ridge with nearly parallel sides or at most slightly constricted (Fig. 50H). Hind femora reddish brown or green: **rhachycoryphus**

FEMALES

Note: Females are very difficult to distinguish. Distributions and associations with males are often helpful in making determinations.

1. Forewings distinctly overlapping dorsally, usually longer than pronotum, and sometimes longer than head plus pronotum:
                                          audax and **aphelocoryphus**
   Forewings not overlapping on dorsum, although occasionally touching: **2**

2. Fastigium of vertex with arcuate groove distant from front margin of depression and posterior margin of pronotal disk usually slightly angulate: **3**
   Fastigium of vertex with arcuate groove at front margin of fastigial depression or at least nearer front than middle; posterior margin of pronotal disk rounded or truncated slightly at center: **5**

3. Inner face of hind femora with black marking. Pronotum very rough, forewings deeply wrinkled (Fig. 51H):   **maculatus**

Inner face of hind femora usually without black marking. Top of
   head, pronotum, and wings not strongly wrinkled or rugose:
                                                              4

4. Rear margin of pronotal disk rounded or at most slightly angu-
   late. Carinulae of fastigium roughly parallel at posterior end.
   Medial area of hind femora usually blackish: **communis**
   Rear margin of pronotal disk angulate, but margin on either side
      of middle slightly concave. Carinulae of fastigium not parallel
      at rear. Medial area of hind femora not black:
                                            thalassinus and **biolleyi**

5. Body pale brown or pale green. Fastigium in most specimens
   convex and with prominent median carinula (Fig. 53C). Accessory carinae well developed, but lateral carinae obsolete.
   Forewings with dark band along upper side of lateral
   field: **salinus**
   Body color reddish brown to dark brown or dark gray. Fastigium
      flat or concave and with at most low carinula. Accessory
      carinae on pronotum usually absent (but present when there is
      strong median pale band). Lateral carinae usually present.
      Forewings not fitting above description: **6**

6. Fastigium relatively narrow; width of fastigium less than 1.25
   times its length (see Fig. 53E for measurement):
                                            chloromerus and **ricei**
   Fastigium relatively wide; width more than 1.35 times length: **7**

7. Top of head sometimes with indistinct median carinula. Width of
   fastigium 1.4 to 1.5 times its length: **rhachycoryphus**
   Top of head without median carinula. Width of fastigium about
      1.5 times its length: **whitei**

Silvitettix communis (**Bruner**)   Pl. 3

DISTRIBUTION. Costa Rica and northwestern Panama.
   RECOGNITION. Males: Variable. Hind femora deep to light purple, or reddish, or pinkish brown. Face and cheeks black. Body with
or without a pale ivory dorsum. Forewings wholly tan to purplish
black. Whole of lateral lobe uniform black or dark olive green. Without accessory carinae.
   Females: Lateral carinae of pronotum weak. Disk of pronotum
flat in profile. Rear margin of pronotum straight, curved or very
slightly angulate. When present, FDI on metazona triangular, never
crossing accessory carinae. Forewings often shorter than pronotum,
bluntly pointed or somewhat rounded at the apex, and usually not

overlapping dorsally. Fastigium with transverse arcuate groove not close to front margin and crossing depression about halfway along its length.

HABITAT. Forest clearings and forest margins.

LIFE CYCLE. Adults have been collected throughout the year.

REFERENCES. Bruner 1904, Jago 1971, Otte and Jago 1979.

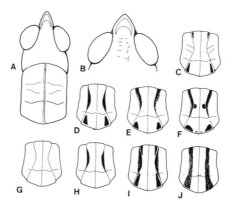

Fig. 49. *Silvitettix communis*. A, male head and pronotum; B, female head; C–J, pronotal variation in females.

Silvitettix biolleyi (**Bruner**)   Pl. 3

DISTRIBUTION. Costa Rica. The species has been collected at Rio Grande, near San Mateo on Rio Surubres, and at Altenas. Jago (1971) suggests that *S. biolleyi* may be an isolated fragment of *S. rhachycoryphus*.

RECOGNITION. Males: Very similar to *S. rhachycoryphus* from Mexico, but differs mainly in having narrower forewings (Fig. 68). Side of body gray to brown, darkening dorsally. Dorsum of body pale tan to yellowish. Hind femora brown. Hind tibiae with a pale band near knees. Venter of abdomen orange. Posterior margin of pronotal disk angulate. Pronotal ridges cut by one sulcus. Rostral length about same as smallest distance between the eyes.

Females: Posterior margin of pronotum broadly rounded. Disk of pronotum decidedly humped (side view). Frontal ridge hardly constricted above ocellus. Accessory carinae weakly present on prozona. Forewings shorter than head plus pronotum, tips reaching back edge of tergite II, and touching each other. Transverse arcuate groove of vertex entirely parallel with front edge of fastigium. Vertex with an indistinct median carinula.

HABITAT. Probably mainly leaf litter in woodlands.

REFERENCES. Bruner 1904, Jago 1971, Jago and Otte 1979.

Silvitettix rhachycoryphus **Jago**   Pl. 3

DISTRIBUTION. States of Chiapas and Oaxaca, Mexico.

RECOGNITION. Males: Very similar to *S. biolleyi* but differing from it in forewing shape (Fig. 50L). Venter of abdomen yellow. Top third of lateral lobes dark chocolate brown. Dorsum of body much lighter than sides. Hind femora rusty brown. Posterior margin of pronotal disk broadly rounded to truncated. Rostral length about 0.9 times smallest distance between eyes. Transverse arcuate groove following front margin of the fastigium. Hind tibiae light brown but sooty distally and below.

Females: Rear margin of pronotum almost straight. Accessory pronotal carinae well developed. Transverse arcuate groove entirely parallel with front edge of fastigium. Vertex with at least a trace of a median carinula. Forewings shorter than pronotum, rounded to slightly pointed, and not overlapping dorsally.

HABITAT. Probably leaf litter in lightly wooded areas.

LIFE CYCLE. The only sample of this species was collected in August 1938.

REFERENCES. Jago 1971, Otte and Jago 1979.

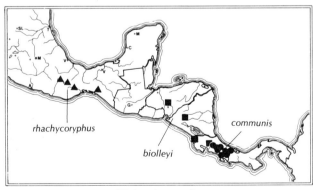

Silvitettix species

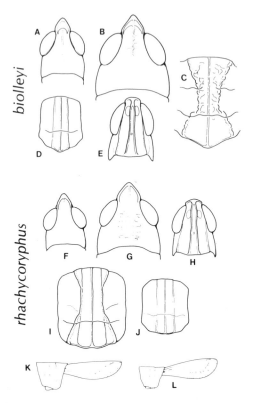

Fig. 50. A–E and K, *Silvitettix biolleyi*. A, male head; B, female head; C, female pronotal disk; D, male pronotum; E, male face; K, male pronotum and forewing. F–J and L, *S. rhachycory-phus*. F, male head; G, female head; H, male face; I, female pronotum; J, male pronotum; L, male pronotum and forewing.

Silvitettix whitei (Hebard)　　　　　　　　　　　Pl. 3

DISTRIBUTION. Oaxaca, Mexico.

RECOGNITION. Males: Similar to *S. rhachycoryphus* but larger. Hind femora 11 mm or more in length. Transverse arcuate groove follows front margin of fastigium. Hind tibiae light brown, becoming sooty apically. Hind femora brown. Body with cream or light brown dorsal stripe and darker brown sides. Abdomen yellow to brown. Length of vertex in front of compound eye same as width, seen from above.

Females: Median carinula of vertex weak and with strong transverse wrinkles more or less obscuring it posteriorly. Large species with back of pronotum humped (side view). Accessory pronotal carinae virtually absent, lateral carinae weak and wrinkled.

HABITAT. Wooded hillsides.

LIFE CYCLE. Adults have been collected in November.

REFERENCES. Hebard 1932, Jago 1971, Jago and Otte 1979.

Silvitettix ricei **Otte and Jago**   Pl. 3

DISTRIBUTION. This species was collected at a single locality in the state of Quintana Roo, Mexico, in July 1975 by R. C. A. Rice. It was found in leaf litter in a tall dry forest twenty-two miles west of Puerto Juarez. Only three males and one female were collected.

RECOGNITION. In general appearance resembling both *S. rhachycoryphus* and *S. biolleyi*, but forewings distinctly shorter than pronotum; lateral pronotal carinae converging in the middle (parallel in *S. rhachycoryphus* and *S. biolleyi*). Body color gray-brown. Dorsum of head, pronotum, and forewings tan, much lighter than side of body. Posterior margin of pronotal disk straight. Venter of abdomen somewhat orange. Top half of medial area of hind femur dark gray. Top third of lateral lobes black, bottom two-thirds gray-brown.

HABITAT. See Distribution.

LIFE CYCLE. The few existing adults were collected in July.

REFERENCES. Otte and Jago 1979.

Silvitettix maculatus **Otte and Jago**   Pl. 3

DISTRIBUTION. Costa Rica and Belize.

RECOGNITION. Males: Dorsum of body pale brown or straw-colored. Sides dark gray to blackish. Forewings shorter than head plus pronotum. Stridulatory file with about 50 teeth. Vertex of head wrinkled. Inner face of hind femora and lower inner carinae with a black marking. Outer face of hind femora slightly greenish, knees black. Hind tibiae blackish, at least in distal half; specimens in Belize entirely black except for a pale annulus near base. Frontal ridge parallel or at most slightly constricted above median ocellus. Transverse arcuate groove crossing fastigium at or near its midpoint. Fastigium decidedly concave. Posterior margin of pronotal disk angulate. Accessory carinae cut by two sulci. Venter of abdomen orange. Body length to ends of femora 19.5 mm.

Females: Dark mark on inner face of hind femora as in males. Top of head rough or wrinkled and usually with a median carinula extending length of head. Arcuate groove near middle of fastigium. Disk of pronotum rough and wrinkled along midline. Hind margin of disk angulate and with left and right margins slightly concave (Fig. 51H). Color quite variable, some mottled with gray and black, some uniformly brownish (various shades), and some blackish on dorsum and brownish on sides. Lateral pronotal carinae usually visible, and all females with dark triangular FDI on metazona.

REFERENCES. Otte and Jago 1979.

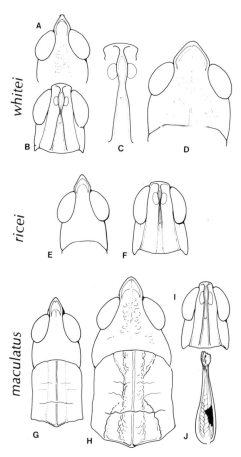

Fig. 51. A–D, *Silvitettix whitei.* A, male head; B, male face; C, frontal ridge of female; D, female head. E–F, *S. ricei.* G–J, *S. maculatus.*

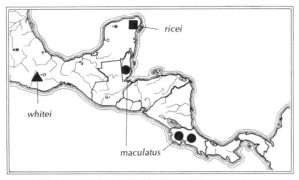

Silvitettix whitei, S. ricei, and *S. maculatus*

Silvitettix audax **Otte and Jago**   Pl. 3

DISTRIBUTION. This species has been collected only at Finca La Paz, near La Reforma, in the province of San Marcos, Guatemala.

RECOGNITION. Males: Side of body with a broad black band from the back of eyes onto forewings and a narrower ivory or yellowish band below the dark band. Lower part of face black, including the lower section of frons, clypeus, labrum, and mouthparts. Hind femora dark blue green. Hind tibiae largely black and with a greenish annulus near base. Venter of abdomen orange.

Females: Forewings longer than pronotum, sometimes as long or longer than head plus pronotum, overlapping medially, and somewhat pointed distally. Vertex of head without a median carinula.

HABITAT. Undergrowth in mountain forests.

LIFE CYCLE. Adults and late instar nymphs were collected by T. H. Hubbell at the beginning of May in 1956.

REFERENCES. Otte and Jago 1979.

Silvitettix aphelocoryphus **Jago**   Pl. 3

DISTRIBUTION. Known only from Rosario and Peña Blanca, 30 miles north of Tegucigalpa in Tegucigalpa Province, Honduras. There it was collected in a cloud forest in the San Juancito Mountains at 4,900–5,100 feet.

RECOGNITION. Males: Top of head, pronotum, and wings much lighter than side of body. Face greenish. Venter of abdomen orange. Lateral lobes distinctly banded: a black band at top, a grayish or

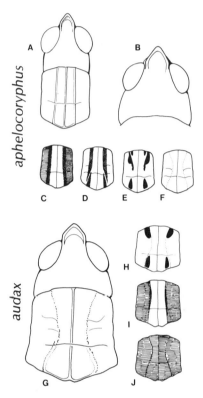

Fig. 52. A–F, *Silvitettix aphelocoryphus*. A, male head and pronotum; B, female head; C–F, female pronotal markings. G–J, *S. audax*. G, female pronotum and head; H–J, female pronotal markings.

dark green band below that, and a yellowish band along bottom. The bottom pale green yellow band extends from face to hind femora. Hind tibiae mostly black, but with a pale band near base. Hind femora brown, knees black. Posterior margin of pronotum slightly angulate.

Females: Forewings always shorter than head plus pronotum, sometimes shorter than pronotum, pointed distally, and usually overlapping dorsally. Rear margin of pronotal disk more rounded than angulate and with median tip flat or slightly concave. Arcuate groove of fastigium very near front margin. Thorax usually with black band along lateral edge of lateral pronotal carinae (as in male). Some females more or less unicolorous brown; others darker on sides of body than on dorsum; others darker on dorsum than on sides; all with a dark band along the side of the abdomen.

HABITAT. Cloud forest near 5,000 feet. (See Distribution).

LIFE CYCLE. The only sample of this species was collected in July.

REFERENCES. Jago 1971, Otte and Jago 1979.

Silvitettix aphelocoryphus and *S. audax*

Silvitettix salinus **(Bruner)** Pl. 4

DISTRIBUTION. Known from a few localities around the Gulf of Tehuantepec, Mexico: near Tuxtla Gutiérrez in Chiapas and near Zanotepec and Ixtepec in Oaxaca.

RECOGNITION. This is the only *Silvitettix* species in which the hind knees are not black. Sexes similar in color. Body pale green or pale brown. Top of head, pronotum, and wings paler than sides. Venter of abdomen yellowish. Hind femora pale green or tan. Hind femora orange in narrowest part. Posterior margin of pronotal disk broadly rounded. Accessory carinae cut by one sulcus.

HABITAT. No records available.

LIFE CYCLE. All adults in the Philadelphia Academy and University of Michigan collections were taken between August and November.

REFERENCES. Bruner 1904, Jago 1971, Otte and Jago 1979.

Silvitettix chloromerus **Jago** Pl. 4

DISTRIBUTION. State of Veracruz, Mexico. The species has been collected near the cities of Córdoba, Atoyac, and Santiago Tuxtla.

RECOGNITION. Males: All legs green. Sides of body darker gray to gray brown. Venter of abdomen yellow to orange. Dorsum of body with a pale gray longitudinal band from the front of the head to the ends of the wings. Hind tibiae green, sometimes turning blackish in distal half. No yellow or pale band near hind knees. Posterior margin of pronotal disk rounded to straight. Accessory carinae diverging slightly toward the back and cut by one sulcus. Fastigium sometimes with a small median ridge. Hind femora with about seventy stridulatory pegs.

Females: Forewings shorter than head, oval, and not overlapping medially. Vertex of head with at least a trace of a median carinula near front. Fastigium with transverse groove at or very near front margin of fastigial depression.

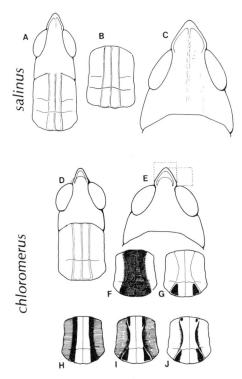

Fig. 53. A–C, *Silvitettix salinus*. A, male head and pronotum; B, male pronotum; C, female head. D–J, *S. chloromerus*. D, male head and pronotum; E, female head, showing method of measuring fastigial dimensions; F–J, pronotal markings in females.

HABITAT. H. R. Roberts collected this species in leaf litter in a coffee plantation growing under the canopy of a natural forest.

LIFE CYCLE. The adults in the Philadelphia Academy and University of Michigan collections were collected in September and November.

REFERENCES. Jago 1971, Otte and Jago 1979.

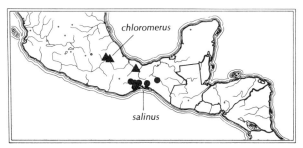

Silvitettix chloromerus and *S. salinus*

Silvitettix thalassinus **Jago**   Pl. 4

DISTRIBUTION. Chiapas, Mexico, to Costa Rica.

RECOGNITION. Males: Side of body grass green or dark olive green, becoming dark gray to black dorsally, or sides entirely dark gray. Dorsum of body pale green or yellow. Front and middle legs green. Hind femora pale yellow or yellow green, but with black knees and a yellowish band encircling narrowest part of femur. Hind tibiae mostly dark gray to black, but with a short yellow band near the knees. Stridulatory file with about 100 pegs. Venter of abdomen orange. Posterior margin of pronotal disk angulate. Accessory pronotal carinae well developed, but lateral carinae obsolete. Rostral length from front of eyes about one and a half times the smallest distance between the eyes.

Females: Disk of pronotum (in side view) with a decided hump on prozona. Rear margin of pronotum angulate. Accessory carinae of pronotum, when present, parallel throughout. Fastigium with transverse arcuate groove far from front margin of fastigial depression, crossing depression halfway along its length. Frontal ridge defined by clear lateral carinae and with a concave median groove. Forewings small, never reaching abdominal tergite V.

REFERENCES. Jago 1971, Otte and Jago 1979.

Silvitettix gorgasi (Hebard)

DISTRIBUTION. Known only from a single specimen taken at Panama City and last collected in 1912.

RECOGNITION. Similar to *S. biolleyi* but differs as follows: lateral pronotal carinae absent or at least very indistinct; disk of pronotum very rough; sides of thorax behind pronotum pitted. Body mostly gray and brown. Top of head, pronotum, and wings pale tan. Lateral margins of disk cut by two sulci. Rostral length about as great as distance between eyes. Posterior margin of pronotal disk slightly angulate. Forewings very narrow. Hind femora light brown. Lateral

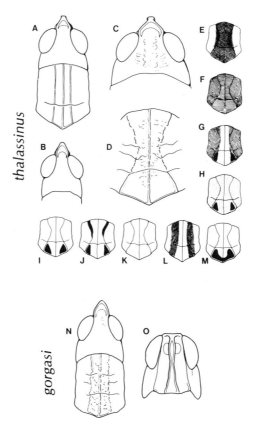

Fig. 54. A–M, *Silvitettix thalassinus*. A, B, male; C, female head; D, female pronotal disk; E–M, female pronotal markings. N, O, *S. gorgasi* males.

lobes slightly darker than pronotal disk. Venter of abdomen pale brown. Top of hind knees brown.

REFERENCES. Hebard 1924a, Jago 1971.

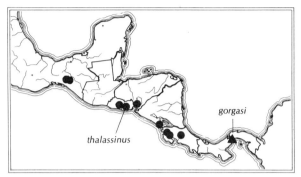

Silvitettix thalassinus and *S. gorgasi*

Orphulella Genus Group (Tribe Orphulellini)

Orphulella Giglio-Tos, *Orphulina* Giglio-Tos, *Dichromorpha* Morse, and the South American genus *Laplatacris* Rehn make up the tribe Orphulellini (Otte 1979).

DISTRIBUTION. The tribe is restricted to North and South America. *Orphulella,* with twenty species, ranges from Canada to Argentina. *Orphulina,* with two species, ranges from Central America to Argentina. *Dichromorpha* has four species and two centers of distribution, in North America from the eastern United States to southern Mexico, and in South America from Bolivia to Argentina. *Laplatacris* is restricted to the pampas from Argentina to extreme southern Brazil.

RECOGNITION. The members of the tribe possess a combination of morphological features that can be used for recognition, but only the first feature listed below is common to all members: Males possess enlarged fore and middle femora. The lateral foveolae of the fastigium are invisible from above. Antennae are usually filiform or at most slightly ensiform. The lateral carinae of the pronotum are usually distinct—in some species they are distinct only on the metazona, and they are either curved inward in the central section or are parallel. The fastigium is concave at least in the anterior part. The forewings vary from being small pads to extending well beyond the ends of the hind femora. The forewings are usually more rounded at the apex than in *Orphula* (Fig. 21C). Males of most species possess stridulatory pegs on the hind femora, but these have been lost secondarily in six species of *Orphulella*.

Members of the Orphulellini are likely to be confused with certain members of the genus *Orphula*. But in *Orphula* the front and middle femora of males are never enlarged, the forewings in both sexes are obliquely truncated, the hind femora of males lack stridulatory pegs, and the hindwings of males always have enlarged cells, whereas only two species of *Orphulella* have enlarged cells.

• Identification of Genera

Orphulella
1. Lateral pronotal carinae constricted in central part (Fig. 20D, F).
2. Posterior FDI usually present
3. Forewings variable in length, usually reaching end of abdomen
4. Shape of hcad and pronotum as in Fig. 55B, C

Orphulina
1. Lateral pronotal carinae moderately constricted in central part (Fig. 20D, F).
2. Posterior FDI usually present
3. Forewings always extending beyond abdomen
4. Shape of head and pronotum as in Fig. 55A

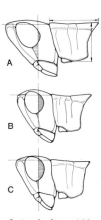

Fig. 55. Comparison of *Orphulina* (A) and *Orphulella* (B, C), showing longer pronotum and more slanting head in *Orphulina*.

Dichromorpha
1. Lateral pronotal carinae parallel or nearly parallel (Fig. 20A)
2. Posterior FDI absent
3. Forewings usually not reaching end of abdomen (but see *prominula*)
4. Shape of head as in Fig. 59

Genus ORPHULELLA Giglio-Tos

DISTRIBUTION. The genus ranges from Canada to Argentina. Seven species are known only from North America, four species are known only from South America, four species range across parts of both continents, and four species are restricted to the Caribbean region. *Orphulella speciosa* and *O. pelidna* in North America and *O. punctata* and *O. concinnula* mainly of South America have extraordinarily wide ranges and may utilize man-made habitats to some extent. *O. punctata*, the most common and widespread species, ranges through much of both continents and the Caribbean region.

RECOGNITION. *Orphulella* differs from *Orphulina* principally in the shape of the head and pronotum (Fig. 55). These differences are by no means pronounced, and some care must be taken to see them. The genus differs from *Dichromorpha* in having converging lateral pronotal carinae (parallel in *Dichromorpha*) and in possessing FDI in the metazona (lacking in *Dichromorpha*). The genus can be further characterized as follows: Lateral foveolae of the fastigium invisible from above. Antennae usually filiform or at most slightly ensiform. Lateral pronotal carinae usually distinct, in some species distinct only on metazona. Lateral carinae either constricted in central part or diverging on metazona (parallel or very nearly parallel in *Dichromorpha*). Disk of pronotum with triangular FDI on metazona (lacking in *Dichromorpha*). Forewings vary from being small pads to extending well beyond ends of hind femora. Males of most species possess stridulatory pegs on hind femora, but these have been lost in six species.

REFERENCES. Gurney 1940, Otte 1979a.

■ Key to Orphulella Species

UNITED STATES AND CANADA
1. Lateral pronotal carinae cut by one sulcus. Forewings usually not reaching ends of hind femora: **speciosa**

Lateral pronotal carinae cut by two or three sulci. Forewings usually extending to or beyond ends of hind femora: **pelidna**

MEXICO

1. Lateral pronotal carinae cut by one sulcus: **2**
   Lateral pronotal carinae cut by two or three sulci: **4**
2. Forewings usually not reaching end of abdomen. Hindwings without enlarged cells. Upper side of lateral lobes black along lateral carinae: **quiroga**
   Forewings always extending to end of abdomen or beyond. Hindwings with enlarged cells: **3**
3. Fastigium with small median carina. Forewings as in Fig. 57:
                                                          tolteca
   Fastigium without small median carina. Forewings as in Fig. 57:
                                                          orizabae
4. Hind femora not spotted (all of United States and much of northern Mexico):   **pelidna**
   Hind femora spotted (central to southern Mexico):   **5**
5. Fastigium proportions as in Fig. 57 (central Mexico):   **aculeata**
   Fastigium proportions as in Fig. 56C (central Mexico to Argentina):   **punctata**

CENTRAL AMERICA

1. Hind femora of males without stridulatory pegs. Posterior margin of pronotum rounded: **2**
   Hind femora of males with stridulatory pegs. Posterior margin of pronotum angulate: **3**
2. Sides of body with postocular dark band: **concinnula**
   Sides of body uniformly dark: **losamatensis**
3. Forewings not reaching end of abdomen (Costa Rica):   **pernix**
   Forewings extending beyond end of abdomen (widespread through Central and South America):   **punctata**

CARIBBEAN

1. Forewings short, not reaching end of abdomen: **3**
   Forewings extending to or beyond end of abdomen: **2**
2. Hind femora in males without stridulatory teeth. Hind femora with prominent spots (Cuba, Ruatan Island):   **scudderi**
   Hind femora in males with stridulatory teeth. Hind femora without prominent spots (widespread):   **punctata**
3. Forewings minute. Back margin of pronotum concave (Fig. 58) (Saint Martin):   **trypha**
   Forewings not as above. Back margin of pronotum convex: **4**
4. Forewings partly overlapping medially. Disk of pronotum not notched at posterior apex. Hind femora of males with stridulatory teeth: **5**

Forewings not overlapping medially. Posterior margin of pronotal disk with median notch. Hind femora of males without
stridulatory teeth (Dominican Republic):   **nesicos**
5. Hind femora black along lower outer margin. Lateral pronotal
carinae distinct throughout their length (Dominican Republic):
                                                          decisa
Hind femora not black or dark along lower outer margin. Lateral
pronotal carinae indistinct in central region (Cuba):
                                                      brachyptera

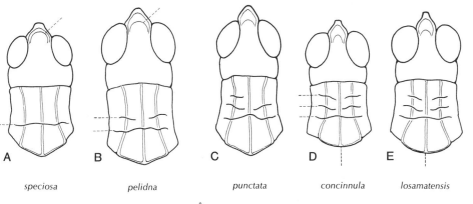

A	B	C	D	E
speciosa	pelidna	punctata	concinnula	losamatensis

Fig. 56. Comparison of head and pronotal features in *Orphulella*
species.

Orphulella pelidna (**Burmeister**)   Pl. 5

DISTRIBUTION. Widespread over North America south to northern Mexico.

RECOGNITION. Most easily confused with *O. speciosa* in the eastern United States, but with the following combination of distinguishing characteristics: Lateral pronotal carinae cut by two and occasionally three sulci; forewings usually extending well beyond hind
femora (in southern California sometimes not reaching end of femora); transverse fastigial depression usually set back from front of
the fastigium; lateral pronotal lobes without the pale curved line of
O. speciosa.

Coloration is highly variable. Some females are mostly green,
others mostly brown. Some individuals have prominent longitudinal
body stripes while others are more unicolorous. A few individuals
have a small pale line on the lower posterior corner of the lateral

lobes. In most males the lateral lobes are mostly dark, except for a pale band in the upper third, and mostly pale in the lower third.

HABITAT. Variable over its range. Abundant in moist grasses in the southeastern United States and common in Atlantic and Gulf coastal marshes. East of the Mississippi inhabiting poor soils from just back of beaches to waste areas such as old fields in upland areas and in open woods (Gurney 1940). In the Great Plains, found in swales and along river valleys (Hebard 1928). In Arizona and southern California found in fields of alfalfa, barley, and grasses and along streams. Along the coast of southern California it is common among the scant grasses of the beach dunes and among low salt-marsh plants.

LIFE CYCLE. Adults are most abundant from July into fall. In Florida adults occur in small numbers throughout the winter, becoming more frequent in February and March.

REFERENCES. Gurney 1940, Brooks 1958, Otte 1970, 1979a.

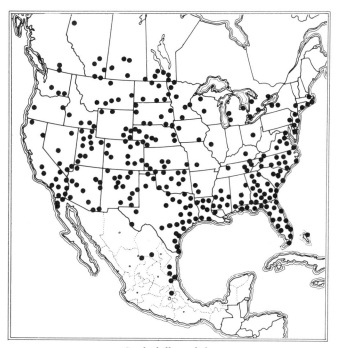

Orphulella pelidna

Orphulella speciosa (**Scudder**)   Pl. 5

DISTRIBUTION. Widespread east of the Rocky Mountains, absent in southeastern states.

RECOGNITION. *O. speciosa* differs from *O. pelidna* as follows: Lateral carinae cut by one sulcus; forewings usually not extending beyond the hind femora; transverse depression usually very near the front of the fastigium. Males often with a pale, oblique line on lateral lobes. Both sexes show green-brown morph variation, with some males mainly greenish on dorsum and others brownish. Females are often brown and with the pale oblique line on the lateral lobe as in males, but a large proportion are pale green with only a trace of dark coloration along lateral carinae. Body length to end of hind femora: 14–21 mm in males, 18–27 mm in females.

HABITAT. Associated with short grasses. Often abundant in upland pastures and semi-waste fields in the northeastern United States; inhabits dry, though not extremely dry, grasslands in the prairie states.

LIFE CYCLE. Adults from May to December in central Texas (most abundant from August through October) and from July through August in New England.

REFERENCES. Gurney 1940, Otte 1979a.

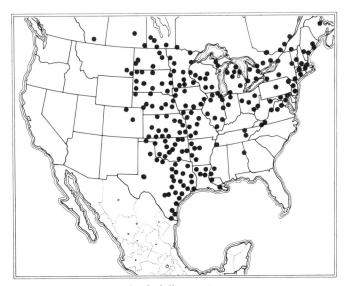

Orphulella speciosa

Orphulella punctata (DeGeer)   Pl. 5

DISTRIBUTION. Central Mexico to Argentina and Caribbean islands.

RECOGNITION. Most easily confused with *O. aculeata* in central
Mexico and with *O. losamatensis* and *O. concinnula* in Central and
South America. The following combination of characteristics distin-
guishes it from other species: hind femora usually with spots along
the lower outer carinula and with one or two larger spots along the
top marginal area (lacking in all above species). Lateral pronotal
carinae prominent throughout and cut by two (sometimes three)
sulci (Fig. 56C); posterior margin of pronotum angulate (rounded in
O. losamatensis and *O. concinnula*). Lateral lobes in males either
with a posterior upward-slanting pale stripe or mostly pale in ventral
half; if mostly pale, the lower margin of the dark area arches up in
the middle section. Hind femora of males with stridulatory pegs
(lacking in *O. losamatensis* and *O. concinnula*). Separable from the
very similar *O. aculeata* mainly by the shape of the fastigium (Fig.
56C).

Variation within populations, especially pronounced in females,
makes identification more difficult. Green females may be confusingly similar to *O. concinnula* but can usually be distinguished by
the lateral carinae, which are distinct in *O. punctata*, low and indistinct in *O. concinnula*, and by the back margin of the pronotal disk,
which is angulate in *O. punctata*, distinctly rounded in *O. concinnula*.

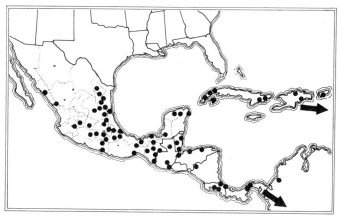

Orphulella punctata

HABITAT. A common tropical species in grassy areas, old fields, pastures, roadsides, and forest edges.

LIFE CYCLE. Adults may be found throughout the year in Central America.

REFERENCES. Otte 1979a.

Orphulella aculeata **Rehn**   Pl. 5

DISTRIBUTION. Central and southwestern Mexico.

RECOGNITION. Very similar to *O. punctata,* from which it can be distinguished by the narrow fastigium; front of fastigium relatively longer and more pointed. A comparison of *O. punctata* from the range of *O. aculeata* shows very little overlap in fastigial width (Fig. 56C). In other respects the two species are quite similar. The following characteristics distinguish *O. aculeata* from *O. tolteca, O. orizabae,* and *O. quiroga* from the same region: Lateral carinae cut by two sulci; hindwings without enlarged cells; hind femora with small spots; lateral carinae cut by two sulci; forewings extending well beyond the hind femora; body length to end of hindwings 17–19 mm in males and 22–25 mm in females. Most males are brown, a few are green on the dorsum and brown on the sides, and still fewer are mostly green. Most females are almost entirely green, but a few are brownish or grayish.

HABITAT. Probably mainly upland grassy tracts.

LIFE CYCLE. Adults have been collected in March, July, August, and September.

REFERENCES. Rehn 1900, Otte 1979a.

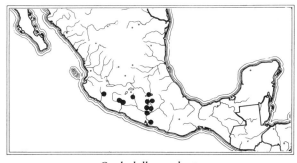

Orphulella aculeata

Orphulella orizabae (McNeill)   Pl. 5

DISTRIBUTION. Highlands of central Mexico.

RECOGNITION. Similar to *O. tolteca*, but differing mainly in hindwing venation and fastigial configuration (Fig. 57). Lateral carinae cut by one sulcus. Hindwing with enlarged cells in the anterior lobe. Fastigium without a small median carina (present in *O. tolteca*). Males and females usually green; males sometimes brown on the side of the body. Hind femora with stridulatory pegs. Wings rarely extending beyond hind femora. Hind femora not spotted as in *O. punctata*.

Almost all individuals are partly green. Males are mostly brown and green on the dorsum. Many males have a pale line ascending to the middle of the lateral lobes from the lower back corner. Some males are mostly dark behind the eyes but have a pale streak running from the back of the eye obliquely down to the pronotum.

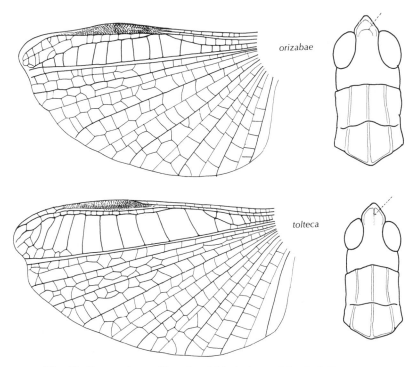

orizabae

tolteca

Fig. 57. Comparison of head and hindwings of *Orphulella orizabae* and *O. tolteca*.

HABITAT. Grasslands and grassy clearings in woodlands.

LIFE CYCLE. Adults in the Philadelphia Academy and University of Michigan collections were all collected in July, August, or September.

REFERENCES. Otte 1979a.

Orphulella orizabae

Orphulella tolteca (**Saussure**)   Pl. 5

DISTRIBUTION. Central Mexico from Mexico City to Guadalajara.

RECOGNITION. Most easily confused with *O. orizabae*, but differing in the shape of the fastigium and the size of the cells on the hindwing (Fig. 57). Lateral carinae cut by one sulcus. Fastigium rather broad and flat and with a small median ridge in the anterior part (*O. orizabae* lacks the median ridge). Males generally dark brown on sides of body and pale brown or green on dorsum. Females usually entirely green but occasionally brown. Male lateral lobes more or less unicolorous and usually dark. Antennae short, slightly ensiform. Males with stridulatory pegs.

Orphulella tolteca

HABITAT. Grasslands, dry hillsides, and among shrubs and oaks.
LIFE CYCLE. Adults in the Philadelphia Academy and University of Michigan collections were taken from May to August.
REFERENCES. Saussure 1861, Otte 1979a.

Orphulella quiroga Otte   Pl. 6

DISTRIBUTION. Central Mexico from Oaxaca to Sinaloa.
RECOGNITION. Forewings usually not reaching the end of the abdomen. Lateral carinae distinct throughout their length and cut by one sulcus. Lateral lobes sooty black along lateral carinae. Hind femora brown to yellowish and without dark spots. Hind knees black in males, variable in females, usually black on the sides. Males almost always brown, but occasionally green on forewings. Females either green or brown. Green females have lateral field of forewing brown. Males with stridulatory teeth on hind femora. Forewing length variable between localities. In most males forewings do not reach end of abdomen; in some males forewings shorter than head plus pronotum. Three males in the Philadelphia Academy collection from Sinaloa have the forewings reaching the end of the abdomen.
HABITAT. Grasslands.
LIFE CYCLE. All adults in the Philadelphia Academy and University of Michigan collections were collected in July, August, and September.
REFERENCES. Otte 1979a.

Orphulella quiroga

Orphulella trypha Otte   Pl. 6

DISTRIBUTION. Saint Martin, Lesser Antilles.

RECOGNITION. Males unknown. Forewings very small and narrow. Lateral carinae distinct throughout and cut by two sulci (Fig. 58). Posterior margin of pronotum concave. Some females are mostly greenish and have a prominent dark stripe along the top of the lateral surface. Other females are mostly straw-colored and lack the dark side band. One female has a broad dark band along the upper side of the body.

REFERENCES. Otte 1979a.

Orphulella nesicos Otte   Pl. 6

DISTRIBUTION. Dominican Republic and Haiti.

RECOGNITION. A small species, males 11–13 mm to end of hind femora, females 16–18 mm. Forewings shorter than pronotum, oval, and not overlapping medially, although almost touching. Lateral pronotal carinae distinct throughout, strongly converging near the middle (Fig. 58). Lateral carinae cut by two or three sulci. Posterior margin of pronotum with a median notch. Males with stridulatory pegs. Hind femora mostly blackish on outer lower marginal area. Top marginal area of hind femora with one very distinct dark marking centrally and usually with two less distinct markings, one on either side of the dark mark, the central mark extending onto medial area. Hind tibiae mostly black, but with a pale band near the proximal end. Lateral field of forewings much darker than dorsal field. Most males are pale brown to straw-colored on dorsum and dark brown to blackish on sides. Lateral lobes usually black along lateral carinae. One male in Philadelphia Academy collection with a pale median stripe from fastigium to end of pronotum.

HABITAT. Philadelphia Academy specimens were collected along with *O. punctata* on a hillside pasture with short grass at about 500 feet.

REFERENCES. Otte 1979a.

Orphulella decisa (Walker)

DISTRIBUTION. Known from a single male collected at Santo Domingo, Dominican Republic.

RECOGNITION. Similar to *O. nesicos* from the same island but differing as follows: Posterior extremity of pronotal disk lacks the
notch of *O. nesicos* (Fig. 58); forewings slightly longer than pronotum (shorter in *O. nesicos*), overlapping medially (no overlap in *O.
nesicos*). Males possess stridulatory pegs (*O. nesicos* lacks pegs).
REFERENCES. Walker 1870, Otte 1979a.

Orphulella brachyptera **Rehn and Hebard**

DISTRIBUTION. Western Cuba.
RECOGNITION. Small species—males about 12 mm to end of hind
femora, females about 17 mm. Forewings (top view) shorter than
head plus pronotum, longer than pronotum. Body color pale brown
or greenish; brown individuals often finely speckled. Forewings
slightly overlapping medially. Upper carina and outer lower carinula
of hind femora with rows of dark spots. Hind tibiae mostly pale
brown, ivory near proximal end. Lateral carinae cut by three sulci.
Lateral carinae indistinct or obsolete between first and third sulci
(Fig. 58). Posterior margin of pronotal disk slightly angulate. Males
with stridulatory pegs.
HABITAT. Sandy pinewoods.
LIFE CYCLE. The few existing specimens were all collected in
September.
REFERENCES. Rehn and Hebard 1938, Otte 1979a.

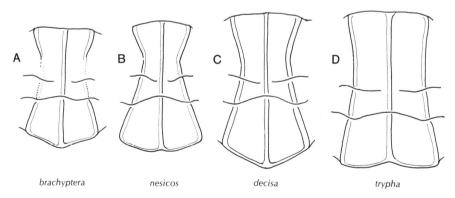

brachyptera   nesicos   decisa   trypha

Fig. 58. Comparison of pronotal disks in four short-winged *Orphulella* species from the Caribbean region.

Orphulella pernix Otte   Pl. 6

DISTRIBUTION. Western Costa Rica.

RECOGNITION. Forewings shorter than head plus pronotum. Lateral carinae rather strongly converging centrally. Lateral carinae distinct throughout and cut by two sulci. Top of hind femora with one to three dark spots, lower outer carinula usually spotted. Hind femora of males with stridulatory pegs. Some males with a pale band across bottom of lateral lobes and a small pale spot near middle of lobes; such males also with a pale median band on dorsum of head, bordered on either side by a longitudinal black stripe; some males with a well-marked pale stripe taking up most of posterior half of episternum 3; some males green on dorsum of pronotum and wings, others pale brown or tan.

LIFE CYCLE. The only known specimens were collected as adults in October 1961.

REFERENCES. Otte 1979a.

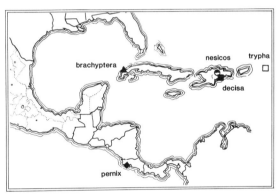

Orphulella pernix, O. brachyptera, O. nesicos,
O. decisa, and *O. trypha*

Orphulella concinnula (Walker)   Pl. 5

DISTRIBUTION. Panama to Brazil and Trinidad.

RECOGNITION. Males with a broad dark postocular band running to ends of forewings (similar but wider band in *Orphulina balloui*). Postocular band in females distinct or indistinct, or obsolete. Posterior margin of pronotum rounded (slightly angulate in *O. punctata* and *Orphulina balloui*). Male femora without stridulatory pegs. Lateral carinae low and almost obsolete, but position clearly indicated. Lateral carinae usually cut by three sulci. Dorsum yellowish or green. Hind femora yellowish or greenish and never with small dark spots. Males often with a thin pale stripe along the wing angle. Forewings usually extending beyond, but sometimes not reaching end, of hind femora, sometimes extending a head's length or more beyond the knees.

HABITAT. Open forest glades, marsh edges, and forest edges. Usually associated with moist grasses.

LIFE CYCLE. A tropical species occurring as adults throughout the year. Seasonal peaks in abundance have not been studied.

REFERENCES. Otte 1979a.

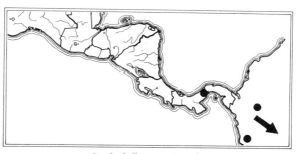

Orphulella concinnula

Orphulella losamatensis Caudell   Pl. 5

DISTRIBUTION. Guatemala, Honduras, Colombia, Ecuador, and Peru.

RECOGNITION. Most easily confused with *O. punctata* and *O. concinnula*. Hind femora without stridulatory pegs (similar to *O. concinnula* and *O. scudderi*). The usual coloration for males is light brown or pale green on top of the body and dark gray brown to black on the side of the body. Females sometimes have similar coloration, but some females are almost as dark on top as on the sides. It also resembles *Orphula azteca,* but the latter lacks thickened forelegs in males and possesses enlarged cells on the hindwings. *O. losamatensis* may be distinguished from all other species by the following combination of characteristics: Posterior margin of pronotal disk broadly rounded (similar to *O. concinnula*). Lateral carinae low, almost obsolete, and usually cut by three sulci. Male femora usually yellowish or pale green and without prominent spots.

LIFE CYCLE. In Guatemala adults have been collected throughout the year.

REFERENCES. Otte 1979a.

Orphulella losamatensis.
Range continues into Colombia, Ecuador, and Peru

Orphulella scudderi (Bolivar)

DISTRIBUTION. Cuba, including Isla de Pinos, possibly Haiti, and Isla de Roatan near Honduras.

RECOGNITION. Males straw-colored or pale green on dorsum and brown to dark brown on sides. Females either mostly green or mostly brown. Forewings of females with a pale streak between the C and Sc veins; in brown females the streak is cream-colored, in green females pale green. Male hind femora without stridulatory pegs. Lateral carinae nearly parallel but narrowing gradually toward

Orphulella scudderi

the front. Lateral carinae cut by two sulci. Fastigium often with a very small median ridge or raised line. Most males in the Philadelphia Academy collection are greenish on dorsum and brown to dark brown on sides. Females are either mostly green or mostly brown.

HABITAT. Rehn and Hebard (1938) noted: "Occasional in short grass along road in low hills and also in more lush meadow grasses." Lutz collected the species among sedges, ferns, and grasses beside a pond and from a weedy xerophytic hill in an unused pasture.

LIFE CYCLE. Adults are present throughout the year.

REFERENCES. Rehn and Hebard 1938, Otte 1979a.

Genus ORPHULINA Giglio-Tos

DISTRIBUTION. The genus includes two species, *O. balloui* Rehn and *O. pulchella* Giglio-Tos and ranges from Panama to Argentina. Only *O. balloui* enters Central America.

RECOGNITION. The genus is very similar to *Orphulella*, differing mainly in the shape of the head and pronotum (Fig. 55).

REFERENCES. Giglio-Tos 1894, Otte 1979a.

Orphulina balloui (Rehn)   Pl. 6

DISTRIBUTION. Honduras to Paraguay.

RECOGNITION. *O. balloui* is most likely to be confused with *Orphulella concinnula,* but has pronotal disk with nearly parallel sides; hind femora with stridulatory pegs (lacking in *Orphulella concinnula*); pale band along bottom of lateral lobes narrow and straight. Lateral carinae cut by two (occasionally three) sulci. Lateral carinae low but distinct throughout and not strongly convergent in the middle section. Hind femora without prominent spots. In many individuals, particularly in the Caribbean region, the side of the body is pale brown, making the bottom pale margin on the lateral lobes indistinct; this pale margin is more distinct in males than in females. Many individuals are mostly dark brown on the side of the body and green on the dorsum. Others are entirely brown, although lighter on dorsum.

HABITAT. Open savannah grasslands and sometimes burned-over areas.

LIFE CYCLE. A tropical species, which probably can be found in the adult stage throughout the year.

REFERENCES. Otte 1979a.

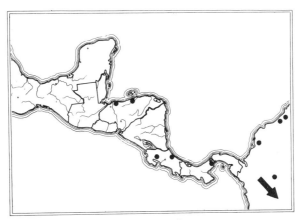

Orphulina balloui

Genus DICHROMORPHA Morse

DISTRIBUTION. Eastern United States to central Mexico and Argentina, Paraguay, and Bolivia. The members of the genus usually inhabit areas where grass cover is thick and not more than knee high, as along the margins of woods, rivers, marshes, and open fields. Three species are known from North America and one from Bolivia, Paraguay, and Argentina. The genus *Clinocephalus* is here treated as a junior synonym of *Dichromorpha*.

RECOGNITION. Lateral foveolae not visible from above. Lateral pronotal carinae well developed, parallel or nearly so (Fig. 93) and cut by one or two sulci. Front and middle femora of males greatly thickened as in *Orphulella*.

REFERENCES. Otte 1979a.

• Identification of Dichromorpha Species

viridis
   1. Lateral pronotal carinae cut by two sulci
   2. Forewings not reaching end of hind femora, but occasionally extending beyond end of abdomen

elegans
   1. Lateral pronotal carinae cut by one sulcus
   2. Forewings not reaching end of hind femora, but occasionally extending beyond end of abdomen
prominula
   1. Lateral pronotal carinae usually cut by one sulcus, occasionally by two
   2. Forewings extending beyond end of abdomen and usually reaching end of hind femora

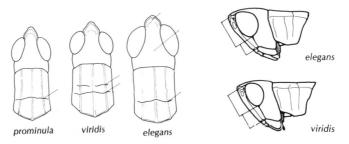

prominula   viridis   elegans   elegans   viridis

Fig. 59. Comparison of head and pronotal features in *Dichromorpha* species.

Dichromorpha viridis (Scudder)   Pl. 6

DISTRIBUTION. Eastern half of United States, eastern and central Mexico.

RECOGNITION. *D. viridis* differs from *D. elegans* as follows: Lateral carinae of the pronotum cut twice (once in *D. elegans*). Head shape differs as shown in Fig. 59. Longest dimension of eye (side view) not more than twice the distance between eye and mandible in males. In females eye length is much less than twice this distance. Hind tibiae with less than twelve outer spines (*D. elegans* has twelve or more spines).

Green-phase males are green on the dorsum and pale brown on the sides; brown-phase males are pale brown on the dorsum and darker brown on the sides. Green-phase females are entirely green; brown-phase females are mostly brown but may be speckled with small dark spots.

In U. S. populations the forewings usually do not reach the end of the abdomen, but in occasional females they do. The ratio of long-winged to short-winged females in the Philadelphia Academy collection varies as follows: Massachusetts, 3:32; Pennsylvania, 5:62; Virginia, 1:68; Indiana, 1:12; Iowa, 0:19; Florida, 3:67; Nebraska, 2:30; Kansas, 2:31; Texas, 3:48; eastern Mexico, 1:19.

The forewings are longer in populations from the western slopes of Mexico than in populations from eastern Mexico and the United States.

HABITAT. Usually inhabiting grasses less than knee high, pastures, roadsides, and margins of woods, lakes, and ponds. In Kansas it is found in brush, timber margins, and other shady places, but not on the open prairie.

BEHAVIOR. Males approach moving individuals stealthily, advancing only when the individual is actually moving. Males pounce onto females without first signaling. Males stridulate only to repel other males (Otte 1970).

LIFE CYCLE. Adults are found from July to September in southern Indiana and throughout the year in Florida. In central Texas adults are most common from August to November.

REFERENCES. Scudder 1862, Morse 1896a, Blatchley 1920, Otte 1970, 1979a.

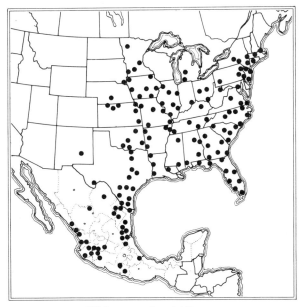

Dichromorpha viridis

Dichromorpha elegans (**Morse**)   Pl. 6

DISTRIBUTION. Gulf and Atlantic coastal regions from New York to Texas.

RECOGNITION. Eyes relatively large; in males the length of the eye is more than twice the distance between the eye and the mandible; in females it is nearly twice this distance. Hind tibiae with twelve or more outer spines. Lateral carinae cut by two sulci. *D. elegans* shows much variation in wing length. In some individuals the forewings are shorter than the head plus pronotum, while in others they may extend to the end of the abdomen. Many individuals have a postocular stripe extending to the end of the abdomen. Females are either mostly green or mostly brown. Many males are green on dorsum and pale brown on the sides. In Florida some males and females possess a dark median dorsal stripe.

HABITAT. Rehn and Hebard (1916) noted: "The species always occurs in moist areas, and in the southeastern U. S. often in woodland, generally among bracken, reeds or grasses, but occasionally on black water-soaked ground covered with low swamp plants. The vicinity of a wet depression or the tangled border of a swampy tract of pine or cyprus is particularly frequented."

LIFE CYCLE. Adult season July into September.

REFERENCES. Morse 1896a, Rehn and Hebard 1916, Blatchley 1920, Otte 1979a.

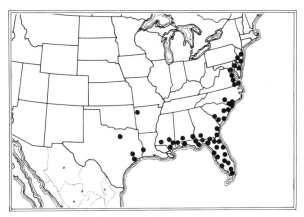

Dichromorpha elegans

Dichromorpha prominula (**Bruner**)   Pl. 6

DISTRIBUTION. Pacific slope of Mexico from Sonora to Guerrero.
RECOGNITION. Likely to be confused with *Orphulella pelidna*.
Lateral carinae spaced more closely than in *D. viridis* (Fig. 59). Most
males possess a dark postocular band which runs onto the forewings; the lower edge of the band is usually not sharp. Males are
either pale green or tan on the dorsum, and all are brownish on the
sides. Forewings in both sexes extend beyond the end of the abdomen and usually reach the ends of the hind femora. Lateral pronotal
carinae are usually cut by one sulcus in Jalisco, Mexico, and northward, but often cut by two sulci in Guerrero. Females, which are
either mostly pale green or pale brown, are more variable in color
than males. Those from Sinaloa are pale brown or pale green and the
majority have a narrow dark line running along the top of the lateral
lobes. In some females from Guerrero, the dorsum is pale brown and
the sides pale green. In a few females from Michoacán the dorsal
field of the wings is pinkish.
HABITAT. Grasslands.
LIFE CYCLE. Adults from August to December in Sinaloa; August
and September in Guerrero.
REFERENCES. Bruner 1904, Hebard 1932, Otte 1979a.

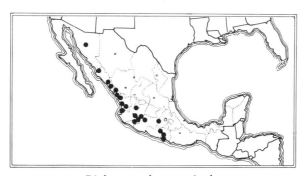

Dichromorpha prominula

Eritettix Genus Group (Tribe Eritettigini)

Three North American genera (*Eritettix, Opeia, Amphitornus*),
one Caribbean genus (*Compsacrella*), and one South American
genus (*Sinipta*) are included in this tribe.

• **Identification of Genera**

Eritettix (Canada to Mexico)
    1. Pronotal disk with FDI
    2. Antennae filiform, club-shaped, or slightly ensiform
    3. Lateral pronotal carinae well developed
    4. Lateral pronotal carinae converging centrally
    5. Hind femora without crossbands
    6. Hind knees light brown
    7. Hind tibiae light brown

Opeia (Canada to Mexico)
    1. Pronotal disk without FDI
    2. Antennae ensiform
    3. Lateral pronotal carinae well developed
    4. Lateral pronotal carinae parallel
    5. Hind femora without crossbands
    6. Hind knees light brown
    7. Hind tibiae light brown

Amphitornus (Canada to Mexico)
    1. Pronotal disk without FDI
    2. Antennae filiform
    3. Lateral pronotal carinae obsolete
    4. Lateral margins of pronotal disk parallel
    5. Hind femora with crossbands (*A. coloradus*) or without (*A. durangus*)
    6. Hind knees black
    7. Hind tibiae blue

Compsacrella (Cuba)
    1. Pronotal disk without FDI
    2. Antennae filiform
    3. Lateral pronotal carinae obsolete
    4. Lateral margins of pronotal disk parallel
    5. Hind femora without crossbands
    6. Hind knees black
    7. Hind tibiae blue

Genus ERITETTIX Bruner

DISTRIBUTION. United States and northern Mexico.

TAXONOMY. As presently understood, the genus includes three species: *E. simplex, E. abortivus* (formerly the only member of the genus *Mesochloa*), and *E. obscurus* (formerly the only member of the genus *Macneillia*). Although *E. variabilis* Bruner was long recognized as a distinct species, it intergrades gradually with *E. simplex* and is here considered a synonym.

RECOGNITION. Antennae variable, usually slightly ensiform, sometimes club-shaped. Top of head with three low, sometimes indistinct longitudinal ridges near the median line: a median ridge and two lateral ridges immediately adjacent (Fig. 29A). In *E. simplex* such ridges also occupy the pronotum. Lateral pronotal carinae distinct and constricted weakly to strongly at the center (nearly parallel in rare specimens). Lateral carinae white and cut by one sulcus or none. Anterior and posterior separation of lateral carinae about equal. Disk of pronotum usually with continuous FDI, but forming triangles when the carinae are strongly constricted centrally (Fig. 29C).

Lateral lobes always with a pale band in the bottom quarter and usually with a narrow white oblique line from the center of the lobe back and down toward the posterior margin. In a rare color morph this line is absent. Hind femora not banded and knees never black. Hind tibiae yellowish or tan. Forewing length variable, rarely extending beyond ends of hind femora.

E. abortivus is somewhat intermediate between *Psoloessa* and *Eritettix*, sometimes possessing faint dark triangular markings on the upper face of the hind femora as in *Psoloessa*, but as in *Eritettix* the antennae are slightly ensiform and the lateral foveolae are invisible from above. *Psoloessa brachyptera* and *P. meridionalis* are also somewhat intermediate between these two genera, having accessory carinulae on the head.

REFERENCES. Scudder 1875a, Bruner 1890, Bruner 1904, Blatchley 1902.

● **Identification of Eritettix Species**

simplex
   1. Widespread, rare in Texas, absent from Florida
   2. Forewings in both sexes extending to end of abdomen
abortivus
   1. Texas, southeastern New Mexico, northeastern Mexico
   2. Forewings in males not reaching end of abdomen, in females shorter than head plus pronotum
obscurus
   1. Florida
   2. Forewings as in *abortivus*

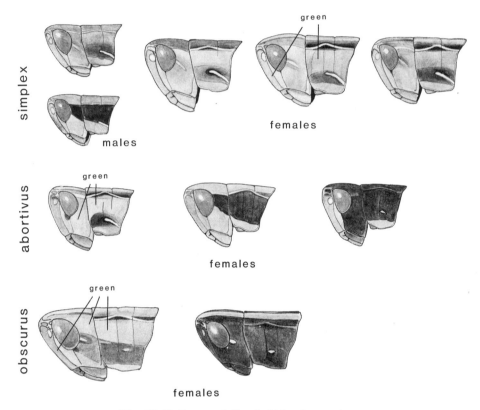

Fig. 60. Pattern variation in *Eritettix* species.

Eritettix simplex (**Scudder**)   Pl. 7

TAXONOMY. *Eritettix variabilis* is here treated as a synonym of *E. simplex*. They were formerly separable mainly on the basis of the antennae (see Variation), but intermediate conditions are common in parts of Colorado.

DISTRIBUTION. There are two main centers, one along the Appalachian Mountains and their eastern slopes and the other in a broad region over the Great Plains from Canada to central Mexico (see Variation).

RECOGNITION. Top of head with three closely spaced low ridges. Disk of pronotum with a median carina and two accessory carinae; the latter are sometimes barely indicated. Lateral carinae white, slightly constricted centrally, and cut by one sulcus. Disk usually with FDI forming dark bands along the lateral carinae and with a pale median band. Male forewings extending to end of abdomen or beyond, occasionally extending slightly beyond hind femora. Female forewings often not quite reaching end of abdomen. In females from Nuevo León, Mexico, and parts of Texas (Brewster County) the forewings are slightly longer than the head plus pronotum. Background coloration variable, some females largely green, others largely brownish or straw-colored. Hind femora often with a dark streak along the upper part of the medial area. Body length to end of forewings 16–20 mm in males, 21–26 mm in females.

VARIATION. Over the greater part of the range, especially in the central and eastern states, the antennae are slightly to moderately club-shaped. Populations from the southwestern states (Arizona and New Mexico, formerly *E. variabilis*) have slightly ensiform antennae. Intermediate conditions occur in Colorado. Bruner (1890) in his original description of *E. variabilis* stated: "very similar in size and general structure to [*E. simplex*] and like that insect also very variable in color. In [*E. variabilis*] the antennae are acuminate instead of clavate, the vertex is narrower between the eyes and the supplementary carinae of the pronotum and occiput are less prominent."

Males and females display two main color patterns (Fig. 60) and within each pattern a gradation of color hues—from green and yellowish to gray and brown. Females are more variable in color than males, and males rarely display traces of green.

HABITAT. In western states, associated with medium- to short-grass prairies. Eastward, inhabiting relict prairies, such as small prairie openings in southern Missouri and hillsides along the eastern bluffs of the Mississippi River in Illinois. Common along eastern slopes of the Appalachian Mountains from Georgia to Connecticut, and in Appalachian region inhabiting shorter grasses of treeless slopes and valleys (Rehn and Hebard 1910).

BEHAVIOR. Pair formation is achieved through loud stridulations by males, and both courtship and aggressive encounters involve stridulation. The calling song is a rapidly delivered sequence of eleven to twenty leg strokes, producing short buzzy sounds (Otte 1970).

In prairie states *E. simplex* is purely a grass and sedge feeder, and at three different localities was found to feed principally on *Bouteloa gracilis*, but it also feeds on ten or more other species (Mulkern et al. 1969).

LIFE CYCLE. Overwinters in late nymphal stages; adults emerge in late spring and early summer, earlier in eastern states. In North

Carolina and Virginia, adults have been collected as early as April. Since adults occur as late as late fall in Oklahoma (Coppock 1962) overwintering may take place in other stages as well. Possibly the species is bivoltine in Oklahoma and Texas.

REFERENCES. Bruner 1890, Rehn and Hebard 1910, Blatchley 1920, Coppock 1962, Mulkern et al. 1969.

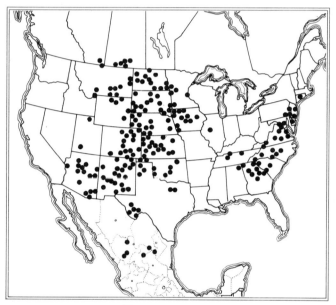

Eritettix simplex

Eritettix abortivus (Bruner)   Pl. 7

TAXONOMY. This species was formerly assigned to the genus *Mesochloa,* of which it was the only member. It is placed in the same genus as *E. simplex* and *E. obscurus* to emphasize its relationships to those species.

DISTRIBUTION. Texas, southeastern New Mexico, and northeastern Mexico.

RECOGNITION. Small and flightless, body length to end of femora 11–15 mm in males, 19–22 mm in females. Top of head with three low longitudinal ridges as in *E. simplex.* Lateral carinae well developed and cut by one sulcus or none. Antennae slightly ensiform. Male forewings about as long as head plus pronotum; female forewings shorter than head plus pronotum, somewhat pointed and slightly overlapping medially. Lateral pronotal carinae distinctly raised, strongly constricted at center; usually not cut by sulci but rarely with one cut. Disk with anterior and posterior triangular FDI. Right and left posterior margins of pronotal disk concave but meeting at an angle. In side view, lateral carinae are bowed upward near center. Hind femora unbanded, but rare individuals have a faint dark marking in the middle of upper face. Coloration variable within populations. Four main patterns are known (Fig. 60), and within each pattern a range of colors may be represented, from reddish or yellowish to gray and green. The genetics of coloration in *E. abortivus* is not known; the basic patterns may have a genetic basis, while color hues are environmentally determined and dependent on the prevailing soil color.

HABITAT. Thinly grassed or bare soils and upland eroded soils from central Texas to New Mexico and northern Mexico. In central Texas abundant on hillsides, roadsides, and in openings in the juniper-oak woodland.

LIFE CYCLE. Overwinters in adult stage; eggs are laid in April and May; hot months are evidently passed in the egg stage. Isely (1937) reported that adults are most frequent from January through April, then disappear for the hot summer months. He found juveniles in September, and adults became common again in October, November, and December.

REFERENCES. Bruner 1890, McNeill 1897b, Scudder 1898, Hart 1906, Isely 1937.

Eritettix obscurus (Scudder)   Pl. 7

TAXONOMY. Originally described under the genus *Chrysochraon*, *E. obscurus* was then for many years placed as the only member of *Macneillia*. Jago assigned it to *Amphitornus*. It is here placed under *Eritettix* for the first time. It differs from *Amphitornus* in possessing lateral carinae that converge centrally.

DISTRIBUTION. Florida peninsula.

RECOGNITION. Vertex of head with three low carinulae. Pronotal disk without accessory carinae except in specimens having a pale dorsal median band. Antennae slightly to moderately ensiform. Lateral carinae distinct, slightly constricted centrally, and not cut by sulci. Right and left posterior margins of pronotal disk slightly concave and together forming a very slight angle. Male forewings longer than head plus pronotum, but not reaching end of abdomen. Female forewings always shorter than head plus pronotum, pointed, and either overlapping or not overlapping medially. Most individuals have a very small white spot near the middle of the lateral lobes, the remains of the oblique pale streak seen in *E. simplex*. Highly variable in color and displaying the following color morphs: (a) uniformly brown or reddish brown; (b) brown to deep reddish brown and with a pale median band on dorsum; (c) side of head and pronotum with a broad green band along the upper side, with a broad, brown band below that, and a narrow ivory band below that, hind femora brown on medial area and green on upper marginal area; and (d) body color mostly brown but with white lateral pronotal carinae and with black triangular FDI next to the carinae (Fig. 60).

HABITAT. Undergrowth and wire-grass in pinewoods. At Live Oak, Florida, *E. obscurus* inhabited "a depression near a sink hole, where the deforested ground was covered with wire-grass and clumps of a dwarf oak growing knee-high" (Rehn and Hebard 1916: 160). At Gainesville it was also found in the undergrowth: "A partiality on the part of the species for low oak growth was noticeable" (Rehn and Hebard 1907: 286).

LIFE CYCLE. Adults in the Philadelphia Academy collection were collected in February, May, July, August, and September.

REFERENCES. Scudder 1877, 1898, Rehn and Hebard 1907, 1912b, 1916, Blatchley 1920.

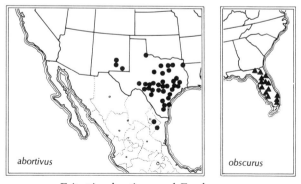

Eritettix abortivus and *E. obscurus*

Genus OPEIA McNeill

DISTRIBUTION. Western half of United States, southern portions of the Canadian prairie provinces, and most of Mexico north of Oaxaca.

RECOGNITION. Most likely to be confused with *Eritettix* and *Amphitornus*. Body color tan, pale green, or pale brown. Pronotal disk never with dark bands or dark triangles along the lateral carinae. Forewings never extending beyond end of hind femora. Antennae ensiform. Lateral foveolae of vertex invisible from above. Fastigium mostly convex and with a strong median ridge. Frontal costa strongly grooved. Lateral pronotal carinae of pronotum with parallel sides and cut by one sulcus. In northern states the disk sometimes widens slightly on the metazona. Lateral lobes variable in pattern (see Fig. 25). Hind femora without crossbands. Hind knees never black on top, but sometimes dark on the sides in *O. atascosa*. Hind tibiae tan or pale brown.

REFERENCES. McNeill 1897a, Bruner 1904, Hebard 1932, 1937.

● **Identification of Opeia Species**

obscura
> Face not so strongly slanted. Front articulation of mandible is anterior to posterior margin of eyes

atascosa
> Face strongly slanted so that front articulation of mandible is posterior to posterior margin of eyes

---

Opeia obscura **(Thomas)**   Pl. 7

DISTRIBUTION. Very widely distributed through western North America.

RECOGNITION. Body color pale brown, yellowish, or pale green. Forewings variable in length, in the United States usually extending to near end of abdomen in both sexes. In central Mexico male forewings about as long as or slightly longer than head plus pronotum; female forewings usually shorter than head plus pronotum, pointed, slightly overlapping medially, and with a black streak (sometimes broken up into a linear series of dark spots) above M vein, and below that often with a pale streak. Hind femora not banded, but usually with a dark line along upper part of medial area. Posterior margin of pronotal disk nearly straight in shorter-winged Mexican specimens, very slightly angulate in longer-winged individuals. Hind tibiae yellowish or pale brown. Differing from *O. atascosa* as follows: Upper front corner of mandible is anterior to a vertical line through hind margin of eye (side view). *Eritettix simplex* individuals having a broad dark band on the lateral lobes may be confused with *O. obscura,* but the former possess three low ridges on the head, and the dark band on lateral lobes becomes distinct from front to back.

HABITAT. *O. obscura* is very widely distributed through the western grasslands and may reach peaks of abundance in the short-grass prairies. In the Great Plains it inhabits dry and sparsely grassed regions where it is closely associated with buffalo grass, *Bouteloua gracilis* (Mulkern et al. 1969). In North Dakota, Hubbell (1922) reported it on the slopes and summits of buttes and in grass, *Artemisia,* and *Opuntia* associations growing on the ridges and gentler slopes in the Badlands. In west Texas, Tinkham (1948) reported it in tall grass on the northern slopes of the Chinati Mountains between 5,000 and 6,000 feet. In Arizona, *O. obscura* sometimes becomes destructive in alfalfa, grains, and Bermuda grass in cultivated areas. It is also common in thicker stands of grass in the Lower Sonoran Zone and in sparse, short grasses of the Upper Sonoran and Lower

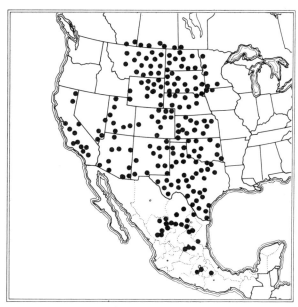

Opeia obscura

Transition zones. In Arizona it feeds on several grasses in the genera *Bouteloua, Sporobolus, Distichlis, Aristida,* and *Muhlenbergia* (Ball et al. 1942, Otte and Joern 1977).

LIFE CYCLE. Adults from April to October in Arizona; from July to end of October in Kansas (Campbell et al. 1974).

REFERENCES. Thomas 1872, Bruner 1904, Hebard 1932, 1937, Ball et al. 1942, Brooks 1958, Mulkern et al. 1969, Joern 1979.

Opeia atascosa **Hebard**   Pl. 7

DISTRIBUTION. Known only from a few localities in Arizona and Mexico. In Arizona it has been collected between 4,500 and 6,500 feet on Atascosa Peak and north of Montana Peak in the Tumacacori Mountains, in the oak belt. A single individual is known from the Santa Rita Mountains. In Mexico it has been collected 12 miles west of Guadalajara, between Zacapu and Quiroga in Michoacán and 5 miles south of Chilchota on the road to Uruapan in Michoacán, and at Tequila in Jalisco.

RECOGNITION. Differs from *O. obscura* mainly in having a more slanted face—the back margin of the eye is anterior to a vertical line through front articulation of mandible. Forewings never reaching end of abdomen. Lateral pronotal lobes strongly or weakly banded; and width of band variable.

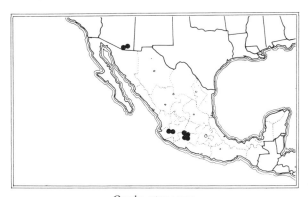

Opeia atascosa

HABITAT. According to Ball et al (1942) *O. atascosa* feeds on tall range grasses such as *Elyonurus barbiculmis* and *Bouteloua curti-*

pendula. Morphologically it appears to be adapted more to a stem-clinging habit than *O. obscura;* its front and middle legs are relatively shorter and its face is more strongly slanted.

LIFE CYCLE. Specimens from Jalisco and Michoacán were collected between July and September. Arizona specimens were collected in September.

REFERENCES. Hebard 1937, Ball et al. 1942.

Genus AMPHITORNUS McNeill

DISTRIBUTION. Western half of United States, southwestern Canada, and parts of Mexico.

RECOGNITION. Vertex of head with a median carinula. Disk of pronotum without lateral carinae and rounding gradually onto the lateral lobes. A pair of accessory carinae run very near and parallel to the median carina; the three carinae are sometimes raised above the remaining disk to form a single low, wide ridge, especially in individuals with a distinct median pale band. The accessory carinae do not extend onto the head as in *Eritettix.* Juncture between disk and lateral lobes cut by three sulci. Back margin of pronotal disk rounded. Top of body usually with parallel light and dark longitudinal bands. Hind knees black. Inner face of hind femora with black bands (upper and outer face also strongly banded in *A. coloradus*). Hind tibiae blue. Side of head with a narrow ivory line beginning at the antennal sockets and curving down and back under the eyes. Lateral lobes with narrow lighter bands along top and bottom and with a white horizontal streak extending from the front to the back margin and running through a broad central dark band. Forewings usually extend beyond the end of the abdomen (except in some populations of *A. coloradus*).

REFERENCES. McNeill 1897b, Otte 1979b.

• Identification of Amphitornus Species

coloradus
    Hind femora banded with black on outer, upper, and inner faces
durangus
    Hind femora not banded on upper and outer faces; medial area
    dark gray to black in upper half to two-thirds

Amphitornus coloradus (**Thomas**)   Pl. 7

DISTRIBUTION. British Columbia to Mexico and ranging through all western states.

RECOGNITION. See generic Recognition for separating this species from other United States Gomphocerinae. Differs from *A. durangus* in having the outer face of the hind femora (excluding knees) strongly banded. A short-winged form, *A. c. saltator* Hebard, has been collected at higher elevations in the Tushar Mountains in Utah, in the Transition Zone of the San Francisco Mountains, and in the juniper-piñon association of the Kaibab Plateau in Arizona. At Pablillo, Nuevo León, Mexico, the forewings of some males and females do not quite reach the end of the abdomen.

HABITAT. Dry grasslands with shorter grasses. Often clings to bunch grasses and is never seen walking about on open ground. In central Colorado, Alexander and Hilliard (1964) reported *A. coloradus* usually occurring below 7,600 feet, but accidentals occasionally are found as high as 11,400 feet. It feeds on grasses and rarely on sedges. At North Platte and Scotts Bluff, Nebraska, it was found to feed principally on *Bouteloua gracilis* and *Stipa comata,* and, at a site in North Dakota, principally on the same two species plus *Poa pratensis.* Small quantities of twelve other grass species were also eaten, as well as trace amounts of two sedges (Mulkern et al. 1969).

BEHAVIOR. The song consists of a series of rather slow notes— zzzz-zzzz-zzzzz-zzzzz—repeated at a rate of approximately three in two seconds.

LIFE CYCLE. Adult season: July to September. Probably egg-overwintering. The season may be longer in Texas and Mexico.

REFERENCES. Thomas 1873a, Hubbell 1922, Hebard 1925b, 1937, Brooks 1958, Strohecker, Middlekauff, and Rentz 1968, Mulkern et al. 1969.

Fig. 61. Pattern variation in *Amphitornus coloradus.*

Amphitornus durangus **Otte**   Pl. 7

DISTRIBUTION. Known only from the state of Durango, Mexico, at 14.1 miles ENE of Llano Grande on Highway 40, and at 38 road miles SW of Durango on Highway 40.

RECOGNITION. Differs from *A. coloradus* in having the outer face of the hind femora unbanded; medial area of hind femora dark gray to black in the upper half to two-thirds and pale along the bottom.

LIFE CYCLE. Adults were collected in August and October.

REFERENCES. Otte 1979b.

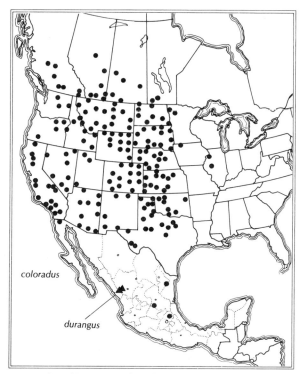

Amphitornus coloradus and *A. durangus*

Genus COMPSACRELLA Rehn and Hebard

DISTRIBUTION. The only species in the genus is known only from Pinar del Río Province, Cuba.

RECOGNITION. See *C. poecila*. Rehn and Hebard believed that this species is related to *Compsacris* from South America, but I agree with Jago (1971) who saw a closer resemblance to *Amphitornus*. The two genera share the following features: Lateral carinae obsolete; pale dorsal band on pronotum slightly raised above the remaining disk; back margin of disk rounded; hind knees black; hind tibiae bluish.

REFERENCES. Rehn and Hebard 1938, Jago 1971.

Compsacrella poecila **Rehn and Hebard**   Pl. 7

DISTRIBUTION. Only four specimens of this species have been collected—two males and a female from 12.5 km S of Pinar del Río, Pinar del Río Province, Cuba, and one male from 13 km S of Pinar del Río. All were collected by F. E. Lutz on September 13 and 14, 1913.

RECOGNITION. Males: Very small—body length to end of hind femora 11.5 mm; hind femur length 7 mm. Hind femora orange with black knees. Hind tibiae blue. Top of body with a pale yellow or cream median band; the pale band is slightly raised above the rest of the disk and bordered on either side by a slightly narrower dark gray to black band. Lateral lobes with a broad horizontal band through the middle. Lateral pronotal carinae at best indistinct. Forewings not reaching end of abdomen; longer than head plus pronotum; with a pale median dorsal band and a yellowish band along lower margin, otherwise dark. Side of head and mandibles yellow-green. Face between preocular ridges and on clypeus and labrum brown, but becoming green on upper side of the frontal costa. Frontal costa grooved except at the top and with nearly parallel sides. Antennae filiform, much longer than head and pronotum. Lateral foveolar area visible from above. Lateral carinal area cut by two sulci.

Females (based on one specimen): Body color similar to male's. Forewings oval, shorter than head, and not overlapping medially. Body length to end of hind femora 17 mm; hind femur length 10 mm. Antennae shorter than head plus pronotum.

HABITAT. This species was collected in grass in sandy pinewoods.

REFERENCES. Rehn and Hebard 1938.

Rhammatocerus Genus Group

The group includes the genus *Rhammatocerus*, which ranges from the United States to Argentina, and *Scyllina*, a South American genus.

Genus RHAMMATOCERUS Saussure

DISTRIBUTION. Southwestern United States to Argentina. *Rhammatocerus* is a comparatively large genus composed of twenty-three nominal species. Only two species, *R. viatorius* and *R. cyanipes*, are known from North America.

RECOGNITION. Relatively large: body length to end of forewings more than 20 mm in males, more than 28 mm in females. Lateral pronotal carinae moderately constricted, indistinct in central part, and cut by three sulci. Disk of pronotum with anterior and posterior FDI. Metazona much longer than prozona. Fastigium of vertex small, concave, and surrounded by broadly rounded surfaces but no distinct ridges. Eyes often with vertical stripes. Hind tibiae orange, red, blue, purple, or a combination of these colors.

REFERENCES. Saussure 1861, Jago 1971.

Rhammatocerus viatorius (**Saussure**)   Pl. 13

DISTRIBUTION. Southwestern Texas and southern Arizona to South America.

RECOGNITION. Forewings extending well beyond hind femora. Body length to end of forewings 29–40 mm in males, 38–51 mm in females. Lateral foveolae absent. Frontal costa convex, rounding gradually onto vertex. Pronotal disk usually with anterior and posterior sets of triangular FDI (when one set is absent, the other is also absent). Face with a black vertical streak descending from bottom of eye. Top of body usually with a pale median line from front of head to rear of pronotum, sometimes extending onto forewings. Side of forewings usually with six or more large dark markings separated by smaller pale bands. Hind femora usually with oblique dark bands in the upper half of the medial area and banded or not banded on upper surface; when banded, upper face has four broad dark bands (including the knee), with broadest band in narrowest part of femur. Hind tibiae variable in color, often red, sometimes bluish distally, and sometimes yellow to orange.

HABITAT. Stony hillsides sparsely covered with grasses and cacti. In Arizona found principally in grasslands in the Baboquivari, Santa Rita, and Tumacacori Mountains. In Texas found on hillsides in the Big Bend region.

LIFE CYCLE. In west Texas nymphs mature in October, and adults are found from late fall through winter until May (Tinkham 1948). In southern Arizona eggs hatch in summer, and adults can be found from early October to late spring. In Mexico and Central America adults occur in all seasons.

REFERENCES. Saussure 1861, Hebard 1924b, Ball et al. 1942, Tinkham 1948.

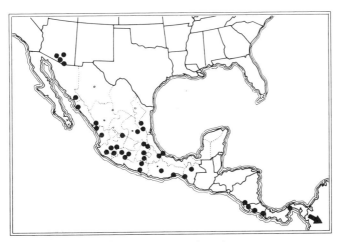

Rhammatocerus viatorius

Rhammatocerus cyanipes (Fabricius)

DISTRIBUTION. Panama, northern South America, and Caribbean islands.

RECOGNITION. Much smaller than *R. viatorius;* body length to end of forewings 20–29 mm in males, 29–38 mm in females. Hind tibiae usually orange proximally and becoming blue or gray in last quarter. Top of the body often with a pale median stripe as in *R. viatorius*.

HABITAT. Probably mainly dry grassy areas and rocky slopes.

LIFE CYCLE. Adults from the Caribbean region in the Philadelphia Academy collection were taken from June through February.
REFERENCES. Fabricius 1775, Hebard 1924b.

Rhammatocerus cyanipes

Aulocara Genus Group

This group is composed of five North American genera (*Aulocara, Ageneotettix, Horesidotes, Eupnigodes,* and *Psoloessa*) and at least the genus *Dociostaurus* from Eurasia. In all genera the face is nearly vertical or at least not strongly slanted, the lateral foveolae are visible from above, and the hind femora are usually banded dorsally.

● **Identification of Genera**

Psoloessa
    1. Hind tibiae brownish or yellowish (*texana*), gray blue (*salina, brachyptera*), or reddish (*delicatula, microptera*)
    2. Lateral carinae distinct (indistinct in *salina*)
    3. Lateral carinae cut by one (all U. S. species) or three sulci
    4. Antennae filiform, brownish
    5. Hind knees brown or pale brown
    6. Head with accessory carinulae (*salina, brachyptera*) or without (*delicatula, texana, microptera*)
    7. Middle dark band on dorsum of hind femora triangular
Eupnigodes
    1. Hind tibiae brownish, tan, or yellowish
    2. Lateral carinae distinct
    3. Lateral carinae cut by three sulci

4. Antennae filiform, pale brown to whitish
5. Hind knees brownish
6. Head without accessory carinulae
7. Middle dark band on dorsum of hind femora triangular

Horesidotes
1. Hind tibiae brownish, tan, or bluish
2. Lateral carinae distinct
3. Lateral carinae cut by one sulcus
4. Antennae slightly ensiform
5. Hind knees brownish
6. Head with accessory carinulae (indistinct in *deiradonotus*)
7. Middle dark band on dorsum of hind femora indistinct or triangular

Ageneotettix
1. Hind tibiae orange or red
2. Lateral carinae indistinct
3. Lateral carinae cut by three sulci
4. Antennae filiform, pale brown or whitish
5. Hind knees black
6. Head without accessory carinulae
7. Middle dark band on dorsum of hind femora triangular (*deorum*) or not triangular (*brevipennis, salutator*)

Aulocara
1. Hind tibiae blue or blue gray
2. Lateral carinae indistinct and strongly converging in central region
3. Lateral carinae cut by three sulci
4. Antennae filiform, long, and black
5. Hind knees black
6. Head without accessory carinulae
7. Middle dark band on dorsum of hind femora not triangular

Genus PSOLOESSA Scudder

DISTRIBUTION. Oaxaca, Mexico, to Canada. In the United States the genus is restricted to the western states.

RELATIONSHIP. The genus is most similar to *Eupnigodes* and *Horesidotes* and somewhat less similar to *Ageneotettix* and *Eritettix*. Rehn (1940) placed the former *Stirapleura salina* (but not the related *Stirapleura brachyptera*) under the genus *Scyllina*. This genus has since been synonymized by Jago (1971) under *Rhammatocerus*. In my view, *salina* and *brachyptera* are quite similar to the members of *Psoloessa,* and I have included them in this genus. They are small, have similar pronotal patterns, possess the black triangle on the dorsum of the hind femora and lack the smoothly rounded forehead of *Rhammatocerus.*

RECOGNITION. In the southwestern United States, *Psoloessa* species are most likely to be confused with *Eupnigodes, Ageneotettix,* and *Horesidotes.* Each genus possesses dark triangular marks on top of the femora (although not all *Ageneotettix* species have them, and the marks may be quite indistinct in *Horesidotes*). In U. S. *Psoloessa,* the lateral carinae are cut by one sulcus (three in *Eupnigodes* and *Ageneotettix*) and are distinct (indistinct in *Ageneotettix*); the hind knees are not black (black in *Ageneotettix*), and the antennae are not pale (pale in one *Ageneotettix* and both *Eupnigodes*).

In Mexico, *Psoloessa* may be distinguished from *Ageneotettix* by the above characteristics and from other genera mainly by triangular marks on the hind femora, constricted lateral pronotal carinae, visible lateral foveolae, and filiform antennae.

REFERENCES. Scudder 1875a, 1876a, Rehn 1942.

• **Identification of Psoloessa Species**

texana (United States to south-central Mexico)
  1. Lateral carinae distinct, cut by one sulcus (United States), occasionally by two or three sulci (Mexico)
  2. Frontal ridge flat in cross section (not grooved or convex)
  3. Hind tibiae pale brown or gray (rarely bluish)
  4. Forewings extend beyond abdomen
  5. Medial area of hind femora without oblique dark streak
delicatula (United States and Canada)
  1. Lateral carinae distinct, cut by one sulcus
  2. Frontal ridge shallowly grooved
  3. Hind tibiae at least partly orange or reddish
  4. Forewings extend beyond abdomen
  5. Lateral pronotal lobes with broad oblique ridge or shoulder; medial area of hind femora with oblique dark streak
brachyptera (Puebla and Oaxaca, Mexico)
  1. Lateral carinae distinct, cut by one sulcus
  2. Frontal ridge flat in cross section
  3. Hind tibiae brownish
  4. Forewings not reaching end of abdomen, but longer than head plus pronotum; forewings overlapping dorsally
  5. Dorsum of head with three low carinulae behind fastigium
microptera (Coahuila and Nuevo León, Mexico)

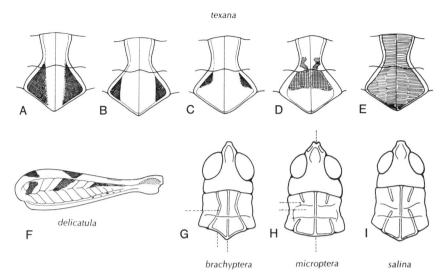

Fig. 62. *Psoloessa*. A–E, variation in *P. texana* female pronotal patterns; F, *P. delicatula* hind femur; G. *P. brachyptera* female; H, *P. microptera* female; I, *P. salina* female.

1. Lateral carinae indistinct in central part and cut by two or three sulci
2. Frontal ridge concave in cross section
3. Hind tibiae partly orange
4. Forewings oval, shorter than head plus pronotum, and not overlapping dorsally
5. Hind margin of pronotum straight.

salina (central Mexico)

1. Lateral carinae indistinct in central part and cut by two or three sulci
2. Frontal ridge convex in cross section
3. Hind tibiae gray to blue
4. Forewings not usually reaching end of abdomen, longer than head plus pronotum
5. Lateral lobes with curved pale band on upper front side

Psoloessa texana Scudder   Pl. 9

DISTRIBUTION. Nebraska and California to Gulf of Tehuantepec, Mexico.

RECOGNITION. Hind tibiae tan, gray, or bluish. Forewing extend-

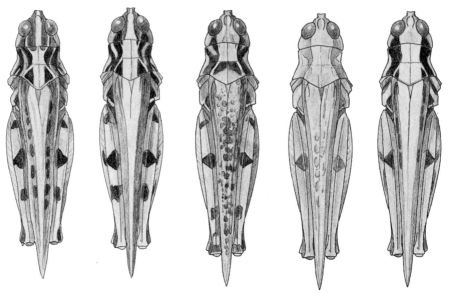

Fig. 63. Pattern variation in *Psoloessa texana* females.

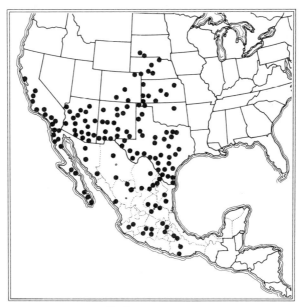

Psoloessa texana

ing beyond end of abdomen and frequently beyond hind femora. Lateral pronotal carinae distinct and cut by one sulcus. Some individuals from central Mexico have carinae cut by two or three sulci. Body length to end of forewings 13–18 mm in males, 17–25 mm in females. Some pronotal patterns are shown in Fig. 63.

HABITAT. Bare open ground with scanty vegetation. In California it has been collected on coastal dunes near Alamitos Bay and Miramar. In Arizona *P. texana* is one of the most ubiquitous and abundant desert-floor species, and ranges upward to 8,000 feet. In Texas it is abundant in grasslands (Tinkham 1948). Although *P. texana* usually frequents bare ground, it retreats to low vegetation and shade during the heat of the day (Otte and Joern 1977).

LIFE CYCLE. Adults from midsummer to fall.

REFERENCES. Rehn 1942, Ball et al. 1942, Tinkham 1948, Otte 1970, Otte and Joern 1977.

Psoloessa delicatula (**Scudder**)   Pl. 9

DISTRIBUTION. Western states, prairie provinces of Canada, and north central Mexico.

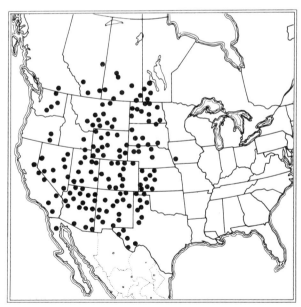

Psoloessa delicatula

RECOGNITION. Hind tibiae orange, sometimes a faint orange tinge confined to the distal extremity. Frontal ridge grooved. Lateral pronotal lobe with a broad rounded ridge or hump beginning at middle front edge of lobe and slanting up to posterior end of lateral carinae. Lateral carinae deeply cut by a single sulcus and strongly depressed in vicinity of cut. Forewings always extending beyond abdomen. Body length to end of forewing 18.0–20.0 mm in males, 16.0–27.2 mm in females. Triangular dark marking on top of hind femora usually connected to an oblique dark streak on medial area (Fig. 62).

HABITAT. Grasslands of the Great Plains and Rocky Mountain region. Usually associated with very short grasses and thinly grassy soils. In northern states it lives at lower elevations, but in Arizona it ranges from 4,000 to 9,000 feet.

LIFE CYCLE. Probably overwinters in nymphal state; adults abundant in spring and early summer and gone by midsummer.

REFERENCES. Scudder 1976a, Rehn 1942, Brooks 1958.

Psoloessa brachyptera (**Bruner**)   Pl. 9

DISTRIBUTION. States of Puebla and Oaxaca, Mexico.

RECOGNITION. Very small, usually short-winged. Body length to end of hind femora 10.5–12.5 mm in males, 13.9–15.5 mm in females. Top of head with three low carinae extending back from the fastigium, a median carinae and two lateral ones, which are extensions of the fastigial carinae (lacking in *P. microptera*). Lateral pronotal carinae distinct and cut by one sulcus (cut twice in *P. microptera*). Both posterior margins of pronotal disk slightly concave and together forming an angle. Forewings variable in length; in most individuals they are somewhat pointed, overlapping medially, and not reaching the end of the abdomen. In most males the forewings are somewhat longer than head plus pronotum; in females they are as long as head plus pronotum or shorter. Forewings in a few males and females collected in Oaxaca and Puebla extend to end of abdomen or beyond. Coloration is highly variable; most individuals with ivory lateral pronotal carinae, some with a median pale band along dorsum and with dark triangular FDI on metazona. Some individuals are mostly straw-colored on dorsum and very dark on sides. Some individuals have lateral lobes with four horizontal bands; in others the lateral lobes are black on the upper side and pale below. Background coloration varies from dark gray or brown to yellowish and pale green.

HABITAT. Grassland and scrub.

LIFE CYCLE. Adults have been collected in May, August, September, December, and January.

REFERENCES. Bruner 1904.

Psoloessa microptera **Otte**   Pl. 9

DISTRIBUTION. States of Nueva León and Coahuila, Mexico.

RECOGNITION. Very small, body length to end of hind femora 11.9–13.0 mm in males and 13.9–15.0 mm in females. Forewings shorter than head plus pronotum, oval, and not overlapping medially. Posterior margin of pronotal disk nearly straight (Fig. 62). Lateral pronotal carinae cut by three sulci. Frontal costa of head grooved. Hind tibiae orange. Body color variable; most individuals brown or straw-colored with black markings. Lateral pronotal carinae ivory. Pronotal disk sometimes with triangular FDI on metazona; sometimes entire posterior border is blackish. Most individuals possess an oblique dark streak on medial area of hind femora extending down and forward from dorsal triangular dark mark.

HABITAT. Desert scrub. Probably lives mainly on the ground.

LIFE CYCLE. Adults have been collected in July and August.

REFERENCES. Otte 1979b.

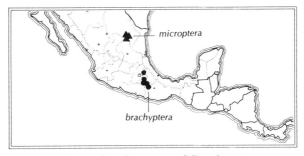

Psoloessa brachyptera and *P. microptera*

Psoloessa salina **(Bruner)**   Pl. 9

DISTRIBUTION. States of Chiapas to Nayarit, Mexico.

TAXONOMY. This species was originally described under the genus *Stirapleura* and then transferred to *Scyllina* by Rehn (1940). I have compared it to *Rhammatocerus* (the senior synonym of *Scyl-*

lina) and *Psoloessa* species and find a greater similarity to the latter. It bridges the gap between *Rhammatocerus* and the more typical *Psoloessa*. Like *Rhammatocerus,* the species possesses poorly defined lateral carinae cut by three sulci, a broad frontal ridge which rounds gradually onto the vertex, anterior and posterior sets of triangular FDI on the pronotal disk, and a broad curved pale band in the upper half of the lateral lobes.

RECOGNITION. Forewings of males usually longer than head plus pronotum, but not extending beyond end of abdomen. Forewings of females usually shorter than head plus pronotum (in a few males collected in Oaxaca, forewings extend beyond abdomen). Lateral pronotal carinae indistinct through most of central region and cut by three sulci; in many individuals the carinae are ivory-colored. Pronotal disk either uniformly pale brown or contrastingly marked with a median pale band, pale lateral carinae, and anterior and posterior sets of triangular FDI, but various intermediate conditions are possible. Lateral lobes variable in color, but in most individuals the lower third is pale with a pale upward-sloping streak beginning at middle front margin and ending along posterior half of lateral carinae; between lower and upper pale bands is a black band which widens from front to back; a small white mark enters the dark band from the lower posterior side; upper front corner of lateral lobes often with a dark triangle. Dorsum of hind femora usually with a smaller dark mark in last quarter and a large triangular mark near middle. Knees of many individuals blackish. Hind tibiae usually gray to blue and with a small pale band near the proximal end. Some females mostly brownish and with only traces of dark patterning on lateral lobes or of a triangular dark mark. Body length to end of femora 13–19.5 mm in males, 17.0–27.0 mm in females.

Psoloessa salina

HABITAT. Weedy areas along roadsides and in openings in lightly wooded area.

LIFE CYCLE. Adults in the Philadelphia Academy collection collected in July, August, and September.

REFERENCES. Bruner 1904, Rehn 1942.

Genus EUPNIGODES McNeill

DISTRIBUTION. California.

RECOGNITION. The genus is most easily confused with *Ageneotettix* and *Psoloessa*, but it differs as follows: Antennae pale (brown in *Psoloessa*); lateral pronotal carinae distinct (indistinct in *Ageneotettix*); lateral carinae cut by three sulci (one sulcus in California *Psoloessa*); frontal ridge grooved (not grooved in *Ageneotettix* and *Psoloessa*); knees not black (black or dark in *Ageneotettix*).

REFERENCES. McNeill 1897a, 1897b, Rehn and Hebard 1909, Rehn 1923, 1927, Thompson and Buxton 1964.

• Identification of Eupnigodes Species

megacephala
1. Forewings not reaching end of abdomen
2. Hind femora usually with oblique dark band on medial area (adjacent to dorsal dark triangle)

sierranus
1. Forewings extending beyond end of abdomen
2. Hind femora usually without oblique dark band on medial area

Eupnigodes megacephala (McNeill)   Pl. 9

DISTRIBUTION. Central California; known only from Alameda, Butte, Contra Costa, Kern, Riverside, San Mateo, and Sonoma counties.

RECOGNITION. Antennae pale yellow or ivory. Forewings not extending much beyond middle of abdomen. Hind femora sometimes with an oblique dark band on the medial area and directly adjacent to dark triangle on upper marginal area. Hind tibiae yellowish to orange. Body length to end of hind femora 13–18 mm in males, 15–21 mm in females.

HABITAT. Salt-grass flats and small water courses in the central valley.

LIFE CYCLE. Adults have been collected in August, September, and October.

REFERENCES. McNeill 1897b, Thompson and Buxton 1964, Strohecker, Middlekauff, and Rentz, 1968.

Eupnigodes sierranus (**Rehn and Hebard**)   Pl. 9

DISTRIBUTION. California: southern portion of the central valley and the mountains south of that valley.

RECOGNITION. Forewings usually extending beyond end of abdomen, rarely beyond ends of hind femora. Antennae whitish. Hind tibiae whitish to pale gray. Some individuals with pale median longitudinal streak on dorsum of body. Medial area of hind femora usually not banded as in *E. megacephala*. Body length to end of hind femora 14–19 mm in males, 18–22 mm in females.

HABITAT. At Raymond, Madera County (altitude 940 feet), specimens were collected on "a hillside stubble field," and at Summit House, Madera County (altitude 2,200 feet) they were collected "on a hillside covered with a scattering growth of oats where the undergrowth consisted of a thick mat of short dry yellow grasses interspersed with very low tar weed" (Strohecker et al. 1968).

LIFE CYCLE. Adults have been collected in July, August, and September.

REFERENCES. Rehn and Hebard 1910, Thompson and Buxton 1964, Strohecker, Middlekauff, and Rentz 1968.

Eupnigodes megacephala and *E. sierranus*

Genus HORESIDOTES Scudder

DISTRIBUTION. Extreme southwestern United States; Baja California, Sonora, and Durango, Mexico.

RECOGNITION. Most similar to *Psoloessa* and *Eritettix*. Lateral foveolae visible from above (invisible in *Eritettix*). Vertex of head with three carinulae (lacking in *Psoloessa* but present in *Eritettix*). Lateral pronotal carinae cut by one sulcus. Antennae slightly ensiform. Lateral lobes usually with a characteristic lower ivory band. Frontal costa slightly concave in cross section at the median ocellus. Top of hind femora sometimes with a dark mark in the central section (mark sometimes triangular as in *Psoloessa*). Medial area of hind femora unbanded.

REFERENCES. Hebard 1931, Otte 1979b.

• **Identification of Horesidotes Species**

cinereus
  1. Forewings usually reaching end of abdomen (except in coastal southern California and Baja California)
  2. Hind knees brownish
  3. Hind tibiae pale brown to grayish
  4. Antennae brownish

deiradonotus
  1. Forewings not reaching end of abdomen, in females shorter than head plus pronotum
  2. Hind knees blackish
  3. Hind tibiae blackish in males, bluish in females, and with pale band near base
  4. Antennae long, blackish

Horesidotes cinereus **Scudder**   Pl. 9

TAXONOMY. Two subspecies have been described: *H. c. cinereus*, in which the forewings extend past the end of the abdomen, and *H. c. saltator*, in which the forewings do not reach the end of the abdomen. The latter is found in southern California and Baja California.

DISTRIBUTION. Southwestern United States, Sonora and Baja California, Mexico.

RECOGNITION. Most likely to be confused with *Psoloessa texana*, but top of head with three low parallel ridges. Antennae slightly ensiform. Lateral lobes of pronotum usually dark in top half, pale in lower half with upper margin of pale area arching up and often nearly white. Posterior margin of epimeron 3 is white. Dorsum usually gray and slightly mottled, but in two less common morphs the dorsum is striped; one has a single pale median band, the other has the median pale band and thin pale lines on the lateral pronotal carinae. Forewing length variable, in some individuals about as long as head plus pronotum, in others just reaching end of the abdomen, shortest in San Diego County. Hind femora yellowish to gray brown, usually faintly mottled; upper marginal area often with a faint triangular marking in central region. In *H. c. saltator,* tops of hind femora often more strikingly banded, with some individuals bearing three dark bands. Hind tibiae pale brown to grayish. In *H. c. cinereus,* body length to end of forewings 15–21 mm in males; 21–28 mm in females.

HABITAT. Desert scrub and chaparral, often inhabiting rocky hillsides.

LIFE CYCLE. Adults in summer and fall.

REFERENCES. Hebard 1931, Otte 1979b.

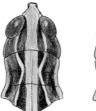

Fig. 64. Pattern variation in *Horesidotes cinereus.*

Horesidotes deiradonotus (Jago)   Pl. 9

DISTRIBUTION. Presently known only from mountain oak-pine woodland along the road between Durango and Mazatlán, Mexico.

RECOGNITION. Sides of body with a black band from back of eye onto forewings. Below this is a cream-colored band, especially pronounced on the lateral lobes, where it bends upward in the middle. Females have a gray band running horizontally through upper dark band (Fig. 115). Disk of pronotum either entirely pale brown or with a median pale band bordered on either side by anterior and posterior triangular FDI. Lateral carinae distinct and cut by one sulcus. Lateral lobes with a definite shoulder along the top. In males, side of head cream or light gray, and face with a dark streak descending along the subocular groove. In females, side of head gray to dark gray brown. Face in both sexes creamy or yellowish between the preocular ridge and the eye and darker between the two preocular

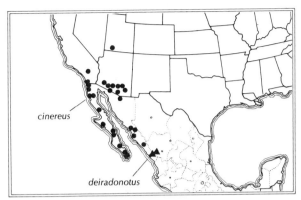

Horesidotes cinereus and *H. deiradonotus*

carinae. Frontal costa broad, very slightly concave below base of the antennae. Lateral foveolae distinct, visible from above. Some individuals, especially those with a narrow dorsal pale band, have three low parallel carinae on the vertex—one median carina and two lateral ones; there are similar ridges on the pronotal disk. Posterior margin of disk rounded to slightly angulate, and left and right margins slightly concave. Forewings in males pale brown on top, dark on the sides, and extending slightly beyond middle of abdomen. Forewings of females slightly shorter than head plus pronotum, somewhat pointed, and touching or slightly overlapping medially. Hind femora yellow along outer lower marginal area and outer lower carinula, becoming dark along upper margin of medial area. Top of femora usually with a small black spot at the center. Hind tibiae

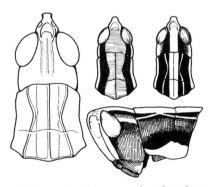

Fig. 65. *Horesidotes deiradonotus,* showing dorsal color variants and accessory carinulae on head.

blackish in males, bluish in females, and with a pale annulus near the base in both sexes. Venter of thorax and abdomen yellow. Body length to end of hind femora 20.0–22.5 mm in males; 26.0–31.5 mm in females; length of hind femora 11.0–12.5 mm in males; 14.5–17.0 mm in females.

HABITAT. Mountain oak-pine woodlands.

LIFE CYCLE. All adults collected in October.

REFERENCES. Jago 1971.

Genus AGENEOTETTIX McNeill

DISTRIBUTION. Western half of North America and northern Mexico.

TAXONOMY. Two species formerly included under the genus *Zapata* Bruner are here included under *Ageneotettix*. In his original description of *Zapata,* Bruner (1904) noted the close similarity between *Zapata* and *Ageneotettix;* Rehn (1927: 214) also put the two genera together under the group Aulocari.

RECOGNITION. Antennae filiform. Lateral foveolae of fastigium visible from above. Lateral pronotal carinae indistinct, with the disk usually rounding gradually onto the lateral lobes. The junction of the disk and the lateral lobes is cut by three sulci. Hind tibiae red. Hind femora with dark knees (sometimes pale on top and dark on the sides). Dorsal surface of hind femora with two or three distinct dark bands (excluding the knees).

REFERENCES. McNeill 1897b, Rehn 1923, Rehn 1927, Bruner 1904.

• Identification of Ageneotettix Species

deorum
1. Central band on dorsum of hind femora more triangular than square
2. Forewings usually extending nearly to end of abdomen (shorter at higher elevations in Arizona and Colorado)
3. Arcuate groove of fastigium near or anterior to middle of fastigium (Fig. 66)
4. Fastigium wider than long
5. Antennae very pale, whitish

brevipennis
1. Central band on dorsum of hind femora not triangular

  2. Forewings oval, shorter than head plus pronotum
  3. Arcuate groove behind middle of fastigium
  4. Width of fastigium nearly equal to length
  5. Antennae pale brown
salutator
  1. Central band on dorsum of hind femora not triangular
  2. Forewings not reaching end of abdomen, but longer than head plus pronotum
  3. Arcuate groove behind middle of fastigium (Fig. 66)
  4. Fastigium much longer than wide
  5. Antennae brown to black

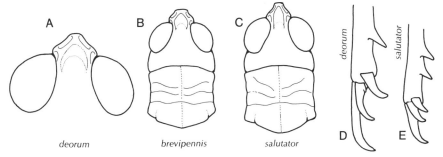

Fig. 66. Males of three *Ageneotettix* species. Comparison of head, pronotal, and tibial features.

Ageneotettix deorum Scudder   Pl. 8

DISTRIBUTION. Longer-winged forms inhabit grasslands from Michigan to California and the prairie provinces of Canada south to north-central Mexico. The brachypterous form formerly known as *A. deorum curtipennis* is found only at higher elevations in Arizona and Colorado. Short-winged forms are common around Prescott, Flagstaff, and Springerville, Arizona.

RECOGNITION. Antennae usually very pale, especially in living males. Hind tibiae red to orange, proximal end black. Medial spur of hind tibiae very long (Fig. 66D). Hind femora with black knees and with three dark marks or bands on the upper marginal area; the front one is sometimes indistinct, and the middle one is triangular. Medial area of hind femora unbanded. Forewings usually somewhat speckled on the lateral field and usually not reaching the end of the abdomen. In shorter-winged forms the forewings are shorter than the head plus pronotum and sometimes do not overlap medially. In rare individuals the forewings extend beyond the end of the abdomen, sometimes even beyond the hind femora. Dorsum of the body sometimes has a pale medial band from the front of the head to the posterior end of the wings. Body length to end of hind femora 11–26 mm in males, 15–26 mm in females.

HABITAT. Associated principally with bare patches of ground in sandy blowouts, short-grass prairies, and desert grasslands (Hubbell 1922, Cantrall 1943, Brooks 1958). It apparently expanded its range eastward into southern Michigan, perhaps between 1940 and 1950 (Cantrall 1943). In the Kansas Flint Hills it feeds mainly on blue grama and Kentucky blue grass (Campbell et al. 1974).

LIFE CYCLE. Adults are found from June to November but appear earlier and disappear later in more southern states. The species overwinters in the egg stage. In Kansas nymphs first appear in early May, and adults are common from late July into August (Campbell et al. 1974).

BEHAVIOR. Pair formation involves much searching and no calling by males. Males wander about, approaching and investigating a

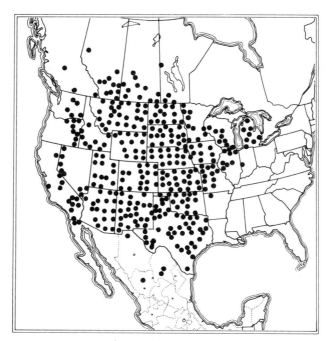

Ageneotettix deorum

variety of moving individuals. They may accidentally approach con-
specific males and females, other grasshopper species, and even
other insects. Solitary males frequently raise and lower their femora
while walking about, apparently signaling their sex to other searching males. Courtship involves only visual signals, such as raising and
lowering the hind femora and the pale antennae (Otte 1970).

REFERENCES. Scudder 1876a, McNeill 1897b, Hebard 1935, Cantrall 1943, Brooks 1958, Otte 1970.

Ageneotettix brevipennis **Bruner**   Pl. 8

TAXONOMY. Bruner (1904) described this species on the basis of a
single female from Lerdo, Durango. Later Rehn (1927) described the
species *Zapata bucculenta* also on the basis of a single female from
Durango City. Hebard (1932: 243) believed the two belonged to the
same species. He wrote: "We believe that *bucculenta* possibly represents a race of *brevipennis* but more probably a synonym." Four
recent series of insects from Sonora, Durango, Chihuahua, and the
Big Bend region have been collected. The considerable variation
even within one locality leads to the conclusion that *Z. bucculenta* is
a synonym of *A. brevipennis*.

DISTRIBUTION. Chihuahuan Desert from the Big Bend region of
Texas to the state of Durango, Mexico.

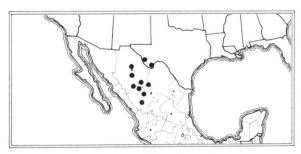

Ageneotettix brevipennis

RECOGNITION. Forewings oval and shorter than the head plus
pronotum. When present, posterior FDI are fused medially to form a
single large triangular mark (Fig. 119). Fastigium broader than in *A.
salutator* (Fig. 66B). Body length to end of hind femora 11.5–
16.2 mm in males, 16.9–21.3 mm in females.

HABITAT. In west Texas *A. brevipennis* is found on the lower slopes of the Chinati Mountains, where it inhabits stony hillsides covered with low scrubby clumps of *Acacia roemeriana* and *Nolina texana* (Tinkham 1948).

LIFE CYCLE. Adult season is late summer and fall; probably egg-overwintering.

REFERENCES. Bruner 1904, Rehn 1927, Hebard 1932, Tinkham 1948.

Ageneotettix salutator (**Rehn**)   Pl. 8

DISTRIBUTION. Southern Arizona to Sinaloa and Baja California, Mexico. A population also believed to belong to this species was dis-

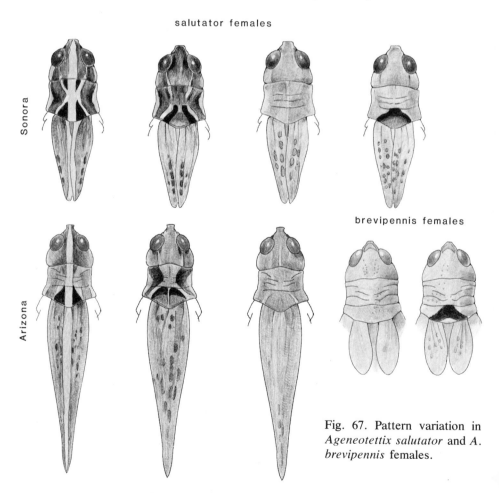

Fig. 67. Pattern variation in *Ageneotettix salutator* and *A. brevipennis* females.

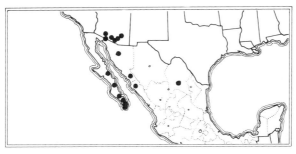

Ageneotettix salutator

covered 33 miles NE of San Pedro de las Colonias in Coahuila along Route 30.

RECOGNITION. Forewings at least as long as head plus pronotum and somewhat pointed (very short and oval in *A. brevipennis*). In Sonora specimens the forewings are short, roughly equaling the head plus pronotum. Antennae brown (very pale in *A. deorum*), frontal ridge wide (Fig. 66C). Coloration is highly variable, but most individuals have a light brown or tan background color. Arizona specimens usually do not have pale stripes at the lateral margins of the pronotal disk as does *A. deorum*. The posterior FDI may be prominent or obsolete; sometimes they join medially to form a single large posterior dark mark (Fig. 67). Dorsum sometimes with a medial ivory band from the front of the head to the ends of the forewings, sometimes uniformly gray or brown. Body length to end of hind femora: 13.0–17.5 mm in males, 17.0–24.0 mm in females.

LIFE CYCLE. Adults in late summer and fall.

REFERENCES. Rehn 1927.

Genus AULOCARA Scudder

DISTRIBUTION. Western North America from central Mexico to southern Canada.

RECOGNITION. Lateral foveolae visible from above. Hind tibiae blue. Hind femora strongly banded; these bands extend onto the medial area as well as onto the inner face of the femora. Knees of males are entirely black, those of females are usually black on the inner face and black on the upper half of the outer face. Lateral pronotal carinae indistinct, strongly converging in central region, and cut by three sulci. Antennae usually long and dark. Lateral lobes usually with a prominent dark marking on most of the upper front quarter; lacking in pale individuals. Front margin of lateral lobes often ivory-colored; lower third and back third pale. Disk of pronotum usually with a pair of prominent ivory stripes along the lateral carinae. Top of folded forewings with a thin, pale median line.

REFERENCES. Scudder 1876b, Bruner 1904, Caudell 1915, Hebard 1935.

• Identification of Aulocara Species

elliotti
1. Forewings extending beyond end of abdomen
2. Dorsal field of forewings usually with pale median stripe along entire length, originating at pronotum
3. Face without small black vertical streak above front articulation of mandible
4. Body length to end of hind femora 16–25 mm in males, 22–35 mm in females

femoratum
1. Forewings usually not reaching end of abdomen (longer than head plus pronotum)
2. Dorsal field of forewings often without pale median stripe; if present, not originating at back margin of pronotum
3. Face often with black vertical streak above front articulation of mandible
4. Body length to end of hind femora 16–23 mm in males, 22–34 mm in females

brevipenne
1. Forewings not reaching end of abdomen (shorter than head plus pronotum)
2. Dorsal field of forewings without pale streak
3. Face sometimes with black streak (as in *femoratum*)
4. Body length to end of hind femora about 13 mm in males, about 20 mm in females

Aulocara elliotti (**Thomas**)   Pl. 8

DISTRIBUTION. Very abundant in the western half of the United States, southern portions of the prairie provinces of Canada, and north-central Mexico.

RECOGNITION. Forewings extend beyond abdomen and usually have a thin pale median stripe running the entire length; in some individuals the band extends forward onto the pronotum and sometimes even onto the head. Disk of pronotum usually with a pair of prominent pale lines converging in the central part and most pronounced on the metazona. Pronotal disk sometimes uniformly straw-colored. Metazona sometimes with two posterior triangular marks separated by a pale median line; when the median line is absent, a single large posterior triangle is formed. Lateral field of forewings usually grayish and with very small dark spots. Body length to end of hind femora 16–25 mm in males, 22–35 in females.

HABITAT. A short-grass prairie species often associated with bare patches of ground. Mulkern et al. (1969) report it feeding almost exclusively on grasses.

BEHAVIOR. Males may be seen wandering about on the ground stridulating from time to time. They usually sing several songs from a small promontory before moving to another spot. Some males may wander about without stridulating and may raise and lower their femora periodically. Courtship usually consists of males silently approaching females and facing them in the region of the abdomen (Otte 1970).

LIFE CYCLE. Adult season in Texas June to September; and in the northern states July to September.

REFERENCES. Thomas 1870, Bruner 1885a, Scudder 1876a, 1876b, Rehn 1927, Bruner 1904, Rehn 1906a, Brooks 1958.

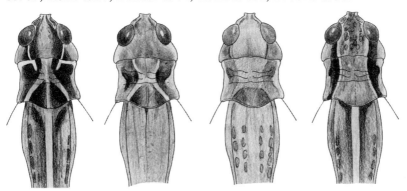

Fig. 68. Pattern variation in *Aulocara elliotti* females.

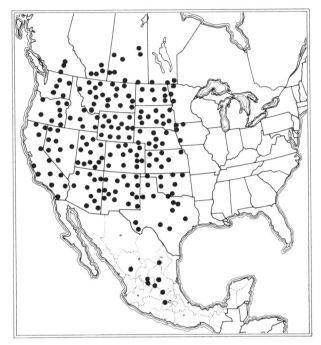

Aulocara elliotti

Aulocara femoratum **(Scudder)**   Pl. 8

DISTRIBUTION. Western North America from Alberta, Canada, to Durango, Mexico. Most common in the mountain states.

RECOGNITION. Very similar to *A. elliotti,* but usually differing as follows: forewings usually not reaching end of the abdomen, but longer than head plus pronotum. In the state of San Luis Potosí, Mexico, forewings sometimes shorter than head plus pronotum. Rare individuals have forewings extending past the abdomen. Top of forewings usually without a pale median line, but when present it usually originates back of the rear margin of the pronotum. Most individuals with a dark vertical streak just above the front articulation of the mandible. Body length to end of hind femora 16–23 mm in males, 22–34 mm in females.

HABITAT. Short-grass prairies and desert grasslands. Usually found where grasses are interspersed with patches of bare ground.

BEHAVIOR. Pairing between the sexes is generally similar to *A. elliotti.* Males wander about on bare patches of ground searching for females. Males stridulate occasionally while courting but usually make conspicuous visual signals with the hind femora (Otte 1970).

LIFE CYCLE. Adults from June to October in Arizona, July to September in Montana.

REFERENCES. Scudder 1876b, Bruner 1904, Rehn 1906a, Rehn 1927, Brooks 1958.

Aulocara brevipenne **Bruner**   Pl. 8

DISTRIBUTION. Known only from the state of Zacatecas, Mexico. Bruner collected it at Comancho, and I collected it on Route 45 about 50 miles NE of Zacatecas City.

TAXONOMY. *A. brevipenne* differs from the more northern species *A. femoratum* mainly in wing length and body size. Possibly the two are geographic varieties of the same species.

RECOGNITION. Very small, body length to end of hind femora about 13 mm in males, about 20 mm in females. Forewings in both sexes shorter than head plus pronotum. Hind margin of pronotal disk very slightly angulate, almost straight. Forewings barely overlapping medially and without a median pale streak on the dorsal field.

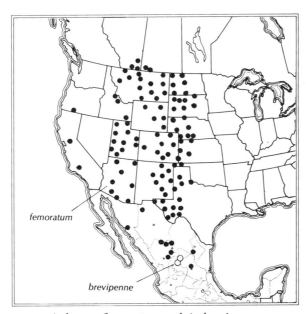

Aulocara femoratum and *A. brevipenne*

HABITAT. I collected this species in an overgrazed creosote desert.

LIFE CYCLE. All adults were collected in October.

REFERENCES. Bruner 1904, Rehn 1927.

Cibolacris Genus Group

This group of grasshoppers consists of desert-inhabiting genera of the southwestern United States: *Cibolacris, Heliaula, Ligurotettix,* and *Xeracris*. The genera were occasionally placed under the Oedipodinae until it was discovered that males possess a femoral stridulatory file. They also have nearly vertical faces; and the pronotal disk lacks lateral carinae and widens strongly from prozona to metazona.

• Identification of Genera

Heliaula
1. Ground-inhabiting
2. Hind tibiae orange
3. Anterior margin of pronotal disk with two bumps
4. Posterior margin of pronotal disk more angulate than rounded (Fig. 69F)

Cibolacris
1. Ground-inhabiting
2. Hind tibiae blue, blue gray, or white (pinkish in *C. weissmani*)
3. Anterior margin of pronotal disk with two bumps
4. Posterior margin of pronotal disk angulate (Figs. 69A, B, C)

Xeracris
1. Ground-inhabiting
2. Hind tibiae yellowish to white
3. Anterior margin of pronotal disk without two bumps
4. Posterior margin of pronotal disk more rounded than angulate (Fig. 69D, E)

Ligurotettix
1. Bush-inhabiting
2. Hind tibiae pale brown or brownish
3. Anterior margin of pronotal disk without two bumps
4. Posterior margin of pronotal disk angulate

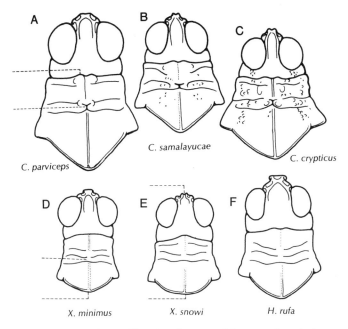

Fig. 69. Comparison of head and pronotal features in *Cibolacris*, *Xeracris*, and *Heliaula* males.

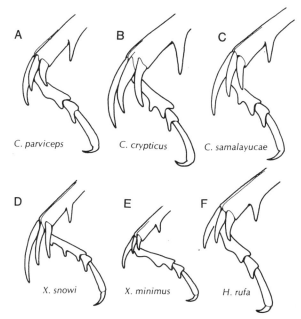

Fig. 70. Comparison of hind tibiae in *Cibolacris*, *Xeracris*, and *Heliaula* males.

Genus HELIAULA Caudell

DISTRIBUTION. See *H. rufa*.

RECOGNITION. Similar to *Cibolacris* and *Aulocara*. Fastigium convex (concave in *Aulocara*). Posterior margin of pronotal disk usually darkened and without two triangular FDI (some *Aulocara elliotti* and *Cibolacris parviceps* individuals also fit this description). Hind tibiae red or orange (blue or gray in *Aulocara* and *Cibolacris*). Lateral carinae indistinct and cut by three sulci. Forewings always extending beyond end of abdomen and usually beyond end of hind femora.

REFERENCES. Scudder 1899b, Caudell 1915.

Heliaula rufa (Scudder)   Pl. 10

DISTRIBUTION. Wyoming and western Nebraska south to Arizona and west Texas.

RECOGNITION. Body color variable; usually pastel hues resembling prevailing ground color. Median carinae barely indicated and cut by one, two, or three sulci. Forewings unicolorous or lightly speckled (never with a pale median streak as in *Aulocara* and never with prominent dark spots as in *Cibolacris parviceps*). Outer face of

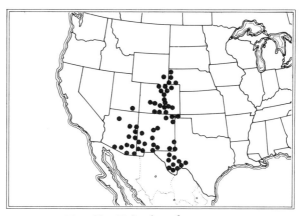

Map 53   *Heliaula rufa*

PLATE 1

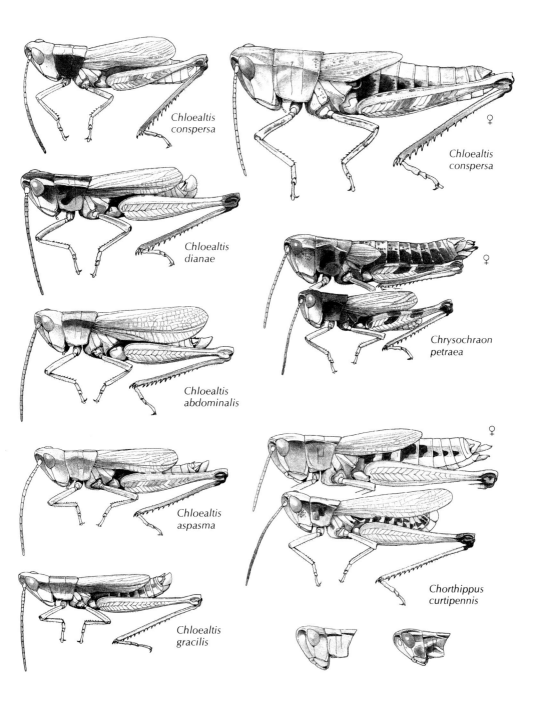

Chloealtis
conspersa

Chloealtis
conspersa ♀

Chloealtis
dianae

Chrysochraon
petraea ♀

Chloealtis
abdominalis

Chloealtis
aspasma

Chorthippus
curtipennis ♀

Chloealtis
gracilis

PLATE 2

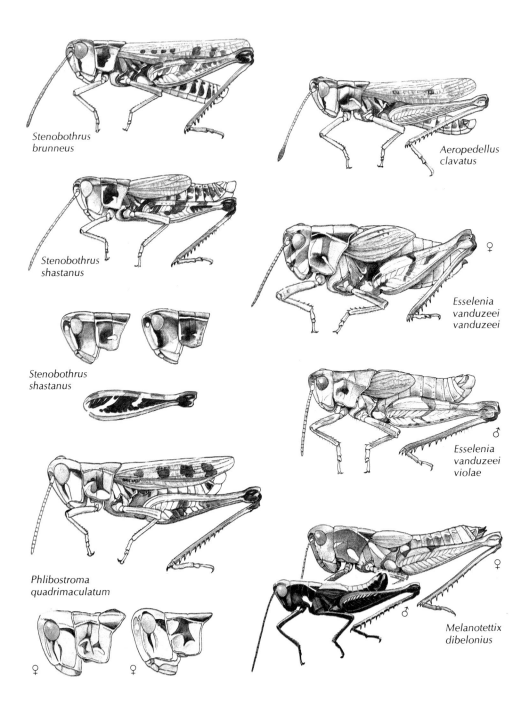

Stenobothrus brunneus

Aeropedellus clavatus

Stenobothrus shastanus

Esselenia vanduzeei vanduzeei ♀

Stenobothrus shastanus

Esselenia vanduzeei violae ♂

Phlibostroma quadrimaculatum ♀ ♀

Melanotettix dibelonius ♀ ♂

PLATE 3

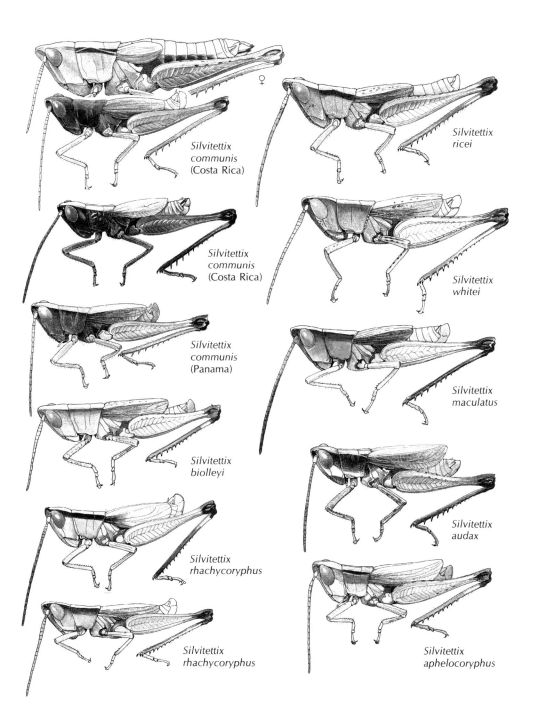

Silvitettix
communis
(Costa Rica)

Silvitettix
ricei

Silvitettix
communis
(Costa Rica)

Silvitettix
whitei

Silvitettix
communis
(Panama)

Silvitettix
maculatus

Silvitettix
biolleyi

Silvitettix
audax

Silvitettix
rhachycoryphus

Silvitettix
rhachycoryphus

Silvitettix
aphelocoryphus

PLATE 4

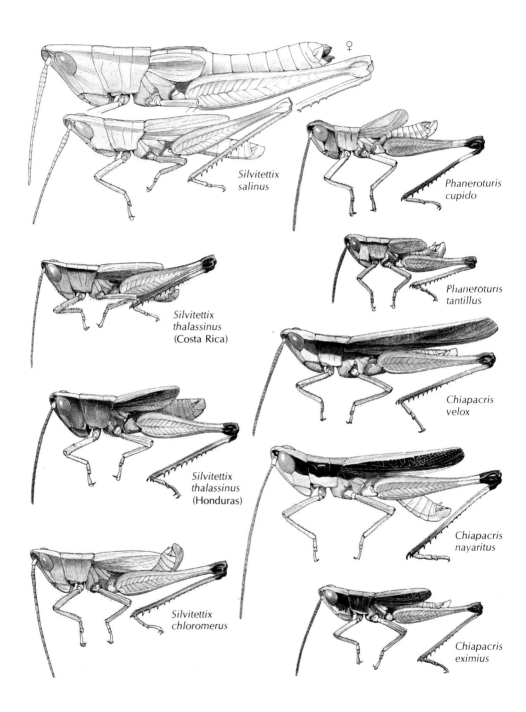

Silvitettix
salinus

Phaneroturis
cupido

Silvitettix
thalassinus
(Costa Rica)

Phaneroturis
tantillus

Silvitettix
thalassinus
(Honduras)

Chiapacris
velox

Silvitettix
chloromerus

Chiapacris
nayaritus

Chiapacris
eximius

PLATE 5

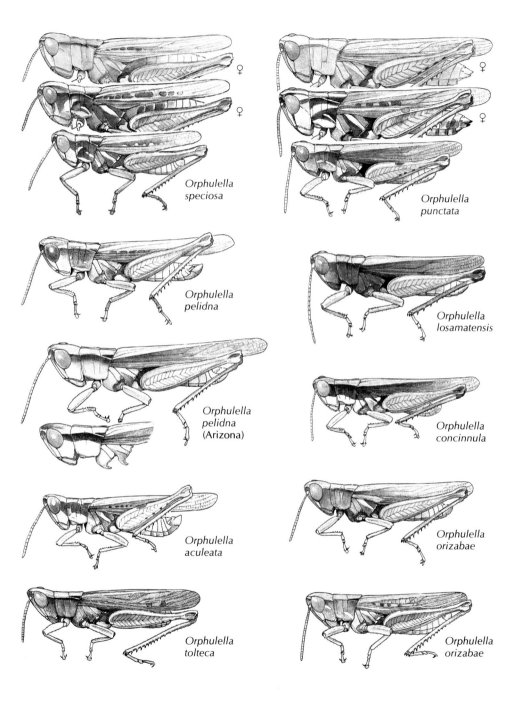

*Orphulella
speciosa*

*Orphulella
punctata*

*Orphulella
pelidna*

*Orphulella
losamatensis*

*Orphulella
pelidna*
(Arizona)

*Orphulella
concinnula*

*Orphulella
aculeata*

*Orphulella
orizabae*

*Orphulella
tolteca*

*Orphulella
orizabae*

PLATE 6

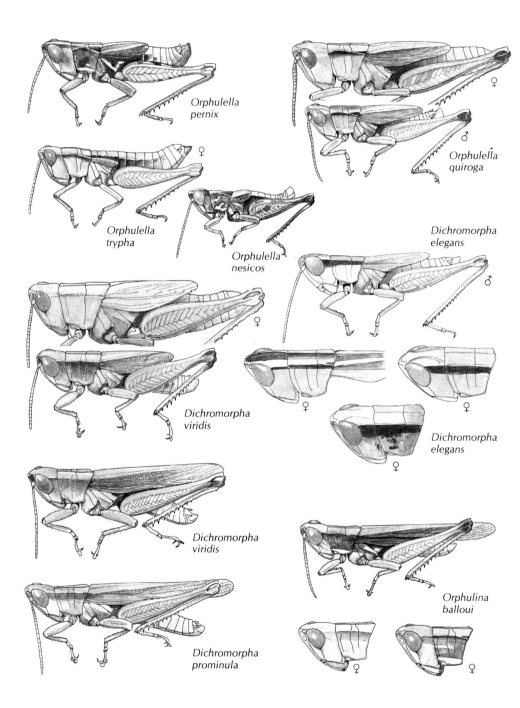

Orphulella
pernix

Orphulella
quiroga ♂

♀

Orphulella
trypha ♀

Orphulella
nesicos

Dichromorpha
elegans

Dichromorpha
elegans ♂

Dichromorpha
viridis ♀

♀

♀

Dichromorpha
elegans ♀

Dichromorpha
viridis

Orphulina
balloui

Dichromorpha
prominula

♀   ♀

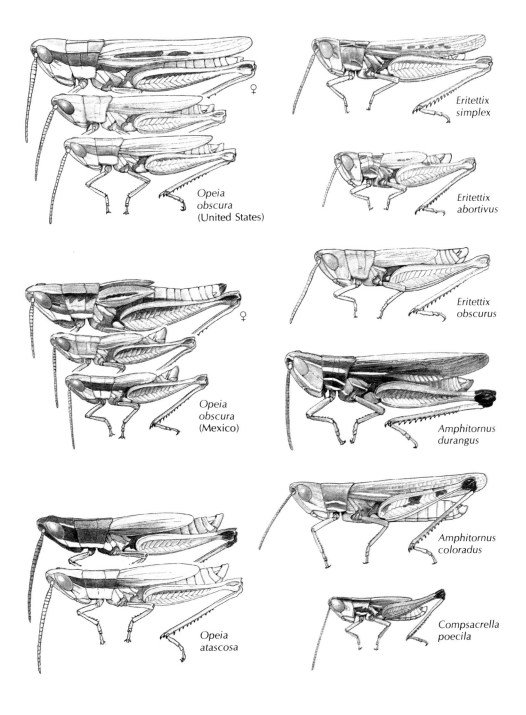

PLATE 7

Opeia
obscura
(United States)

♀

Eritettix
simplex

Eritettix
abortivus

Opeia
obscura
(Mexico)

♀

Eritettix
obscurus

Amphitornus
durangus

Amphitornus
coloradus

Opeia
atascosa

Compsacrella
poecila

PLATE 8

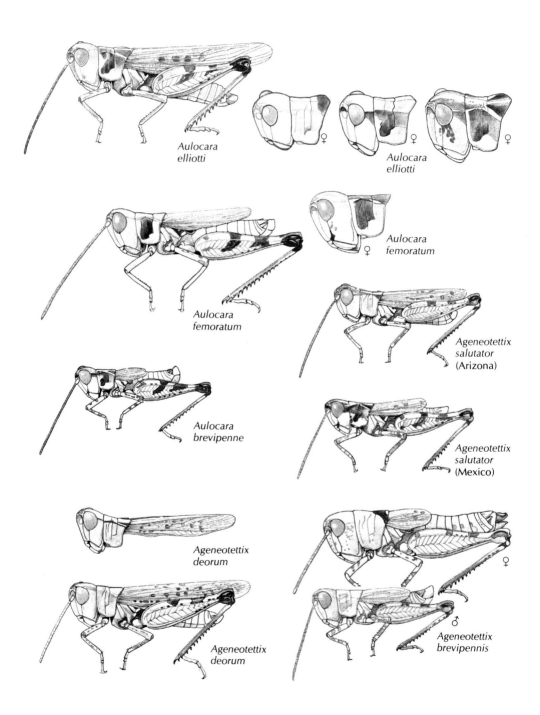

Aulocara
elliotti

Aulocara
elliotti
♀ ♀ ♀

Aulocara
femoratum
♀

Aulocara
femoratum

Ageneotettix
salutator
(Arizona)

Aulocara
brevipenne

Ageneotettix
salutator
(Mexico)

Ageneotettix
deorum

Ageneotettix
deorum

Ageneotettix
brevipennis
♀
♂

PLATE 9

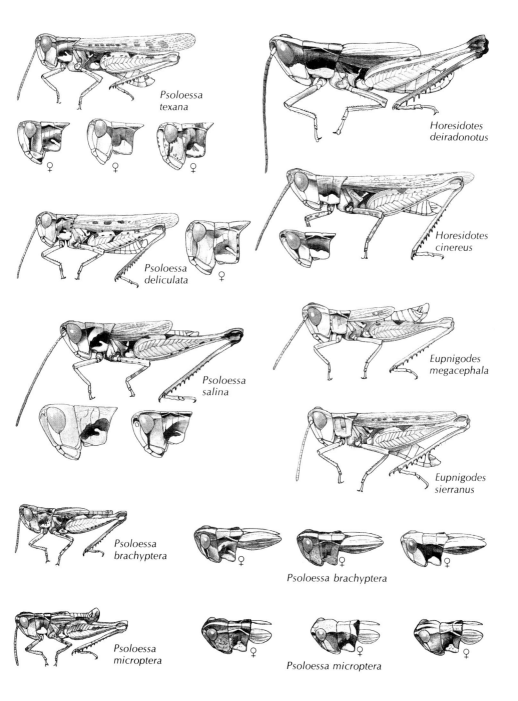

*Psoloessa
texana*

♀ ♀ ♀

*Horesidotes
deiradonotus*

*Psoloessa
deliculata*
♀

*Horesidotes
cinereus*

*Psoloessa
salina*

*Eupnigodes
megacephala*

*Eupnigodes
sierranus*

*Psoloessa
brachyptera*

Psoloessa brachyptera
♀ ♀ ♀

*Psoloessa
microptera*

Psoloessa microptera
♀ ♀ ♀

PLATE 10

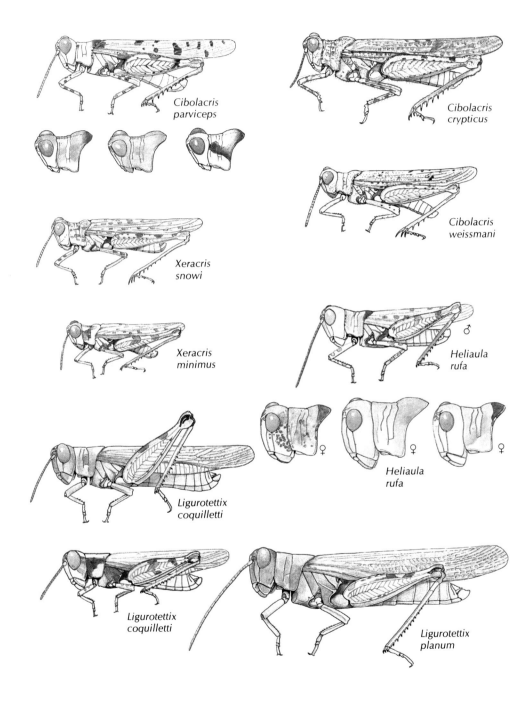

Cibolacris
parviceps

Cibolacris
crypticus

Cibolacris
weissmani

Xeracris
snowi

Xeracris
minimus

Heliaula
rufa ♂

Heliaula
rufa ♀ ♀ ♀

Ligurotettix
coquilletti

Ligurotettix
coquilletti

Ligurotettix
planum

PLATE 11

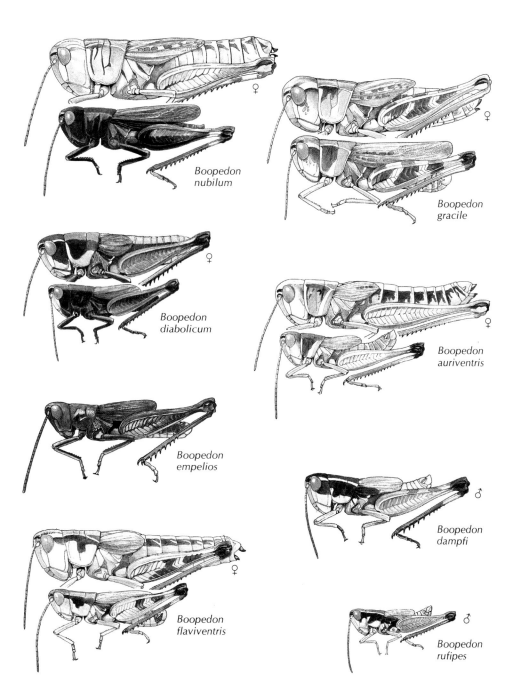

Boopedon nubilum ♀

Boopedon gracile ♀

Boopedon diabolicum ♀

Boopedon auriventris ♀

Boopedon empelios

Boopedon dampfi ♂

Boopedon flaviventris ♀

Boopedon rufipes ♂

PLATE 12

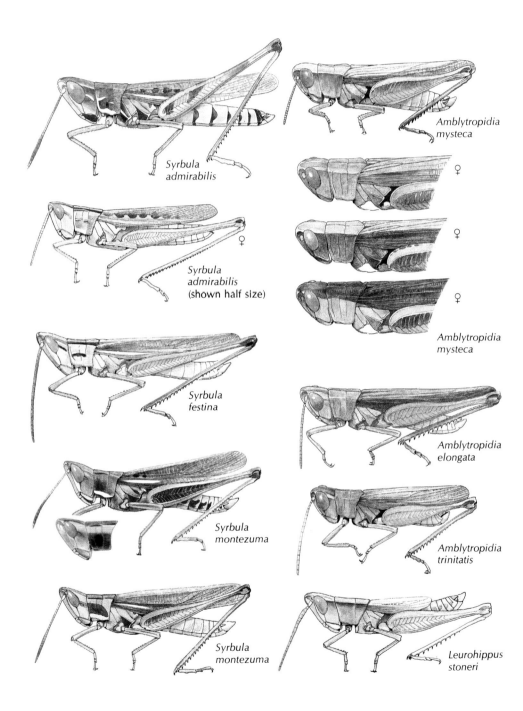

Syrbula
admirabilis

Amblytropidia
mysteca

Syrbula
admirabilis
(shown half size) ♀

♀

♀

♀

Amblytropidia
mysteca

Syrbula
festina

Amblytropidia
elongata

Syrbula
montezuma

Amblytropidia
trinitatis

Syrbula
montezuma

Leurohippus
stoneri

PLATE 13

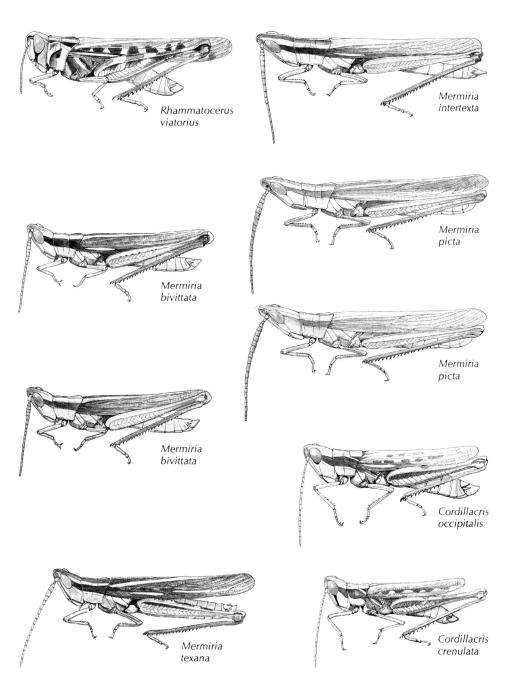

Rhammatocerus
viatorius

Mermiria
intertexta

Mermiria
bivittata

Mermiria
picta

Mermiria
picta

Mermiria
bivittata

Cordillacris
occipitalis

Mermiria
texana

Cordillacris
crenulata

PLATE 14

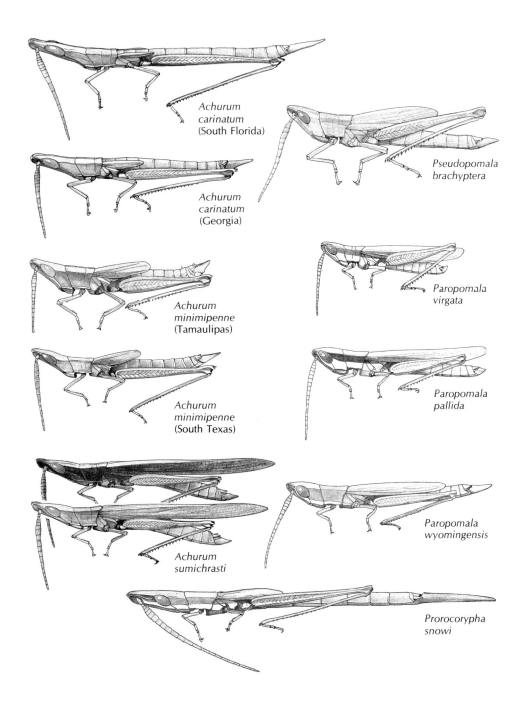

Achurum
carinatum
(South Florida)

Pseudopomala
brachyptera

Achurum
carinatum
(Georgia)

Paropomala
virgata

Achurum
minimipenne
(Tamaulipas)

Achurum
minimipenne
(South Texas)

Paropomala
pallida

Paropomala
wyomingensis

Achurum
sumichrasti

Prorocorypha
snowi

PLATE 15

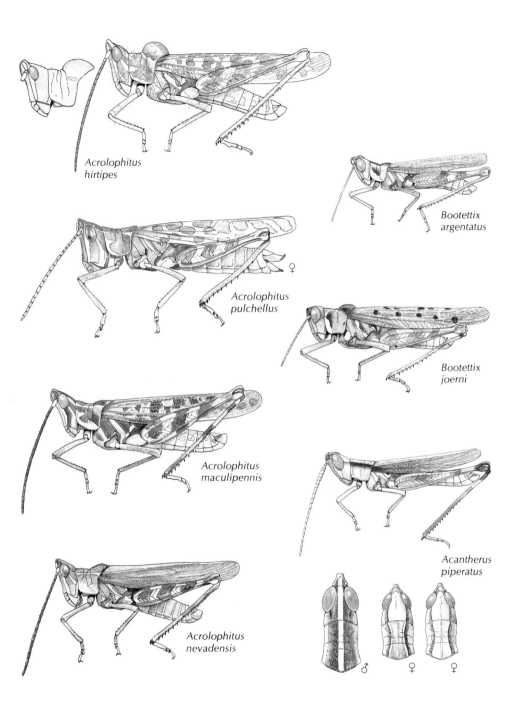

Acrolophitus
hirtipes

Bootettix
argentatus

Acrolophitus
pulchellus

Bootettix
joerni

Acrolophitus
maculipennis

Acantherus
piperatus

Acrolophitus
nevadensis

PLATE 16

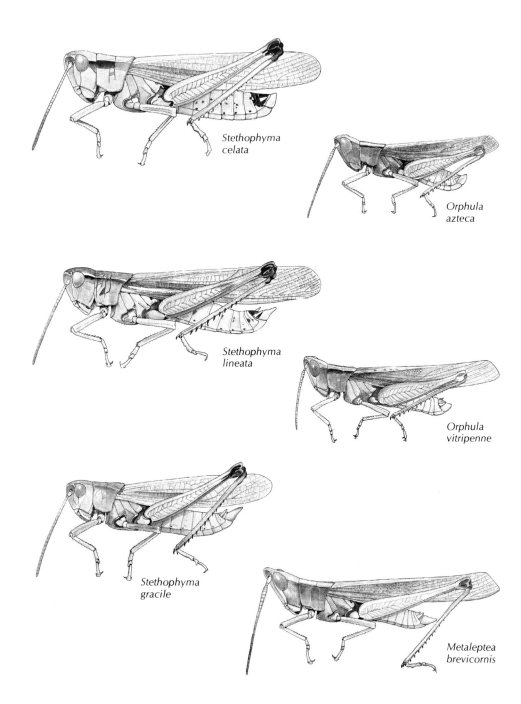

Stethophyma
celata

Orphula
azteca

Stethophyma
lineata

Orphula
vitripenne

Stethophyma
gracile

Metaleptea
brevicornis

hind knees mostly pale, but crescent may be dark. Inner face of femora without dark bands. Dorsum of hind femora with two dark bands anterior to knees; these sometimes extend onto medial area. In very pale individuals such bands are barely visible. Body length to end of forewings 14–20 mm in males, 20–27 mm in females.

HABITAT. Usually gravelly soils, often on hillsides with sparse vegetation. In Arizona, mainly the upper margins of the desert and the Upper Sonoran Zone from 2,000 to 7,000 feet (Ball et al. 1942).

LIFE CYCLE. Adults from June to October.

REFERENCES. Scudder 1899b, Caudell 1915, Ball et al. 1942, Tinkham 1948.

Genus CIBOLACRIS Hebard

DISTRIBUTION. Sonoran, Chihuahuan, and Mojave deserts.

TAXONOMY. This genus presently includes four species, *C. parviceps, C. crypticus, C. samalayucae,* and *C. weissmani.* The genus was long placed in the family Oedipodinae, but because it possesses stridulatory pegs on the inner face of the hind femora, Jago (1971) placed it under the Gomphocerinae. In general appearance it is most similar to *Heliaula* and *Xeracris.*

RECOGNITION. Lateral pronotal carinae obsolete, and disk rounding gradually onto lateral lobes. Lateral margins of disk usually cut by three sulci. Median carina absent, indicated by a line. Differing from *Heliaula* and *Xeracris* as follows: Front margin of disk with two bumps (Fig. 69) (also present in *Heliaula*). Posterior sulcus on pronotum connected at the median line to a small pit or depression which is a remnant of the second transverse sulcus. *Xeracris* and *Heliaula* lack the depression. Hind femora whitish or bluish (reddish in *Heliaula*). Females with stridulatory teeth on hind femora.

REFERENCES. Walker 1870, Hebard 1937, Gurney 1940, Vickery 1969a, Jago 1971.

• Identification of Cibolacris Species

parviceps (widespread)
1. Body color not finely granular but usually rather clearly banded or marked with larger spots
2. Posterior end of pronotal disk usually sooty or darker than remainder of disk

3. Dorsum of hind femora with distinct dark band on narrowest part
4. Pronotal disk relatively smooth (Fig. 69A)
5. Hind tibiae bluish or whitish

crypticus (Sonora and Baja California)
1. Body color sandy and finely granular
2. Posterior end of pronotal disk not darkened
3. Dorsum of hind femora not banded as in *parviceps*
4. Pronotal disk highly rugose (Fig. 69C)
5. Hind tibiae bluish or whitish

samalayucae (Chihuahua)
1. Body color sandy or finely granular
2. Posterior end of pronotal disk not darkened
3. Dorsum of hind femora not banded as in *parviceps*
4. Pronotal disk intermediate between *parviceps* and *crypticus* (Fig. 69B)
5. Hind tibiae whitish or bluish.

weissmani (Baja California)
1. Body color sandy and with *very* fine spots
2. Posterior end of pronotal disk not darkened
3. Dorsum of hind femora not banded as in *parviceps*
4. Pronotal disk not very rough
5. Hind tibiae usually pinkish

Cibolacris parviceps (**Walker**)   Pl. 10

DISTRIBUTION. *C. parviceps* ranges over the entire Sonoran-Mojave Desert and much of the Chihuahuan Desert and chaparral regions of southern California. *C. p. californicus* (Thomas), a wide-bodied form, is known only from Los Angeles, Orange, and San Diego counties, California, where it has been found at Los Angeles, Pasadena, Claremont, Riverside, Anaheim, and San Diego.

RECOGNITION. Pale, pastel-colored insects—yellowish, cream, bluish, reddish, and pinkish individuals are common. The color depends on the prevailing color of the substrate on which they are found. Top of hind femora banded, with the darkest band on the narrowest part; the front of this band is indefinite, but posterior margin is black and sharp. Inner face of hind femora unbanded. Hind tibiae bluish, pale gray, or white. Dorsal field of forewings with one to six

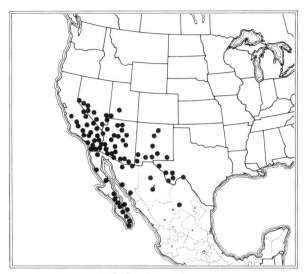

Cibolacris parviceps

small dark crossbands or spots; the most anterior of these is the darkest and is always present. In reddish individuals the dark bands on the forewings and hind femora are often indistinct. Body length to end of forewings 19–25 mm in males, 28–34 mm in females. In the coastal subspecies *C. p. californicus,* the head is very narrow relative to the pronotum, and the top of the pronotum has numerous small tubercles and bumps.

HABITAT. One of the most ubiquitous grasshoppers in the southwest. It lives on the ground, frequently in areas dominated by creosote bushes or desert grasses and seems to prefer stony ground to sandy areas. In the Sonoran Desert one of its major food plants is creosote bush. Other annual herbs are also eaten when available (Otte and Joern 1977). In creosote deserts this species remains cryptically concealed on the desert floor during the daytime, but can be found feeding on bushes at night. Both acoustical and visual signals are used by courting males (Otte 1970).

LIFE CYCLE. Adults are found from late March to November in Arizona. Development depends on the time of spring and summer rainfall. In areas with colder winters, the adult season appears to be shorter.

REFERENCES. Walker 1870, Thomas 1874, Scudder 1897, Rehn and Hebard 1909, Hebard 1937, Otte 1970.

Cibolacris crypticus (Vickery)   Pl. 10

DISTRIBUTION. Baja California and Sonora, Mexico.

RECOGNITION. Body color granular and blending with the prevailing sand color. Pronotum with small tubercles, two larger tubercles on disk in front of the anterior sulcus. Hind margin of pronotal disk not darkened as in *C. parviceps*. Darker spots on dorsal field of forewings either absent or indistinct. Terminal spurs of the hind tibiae relatively long (Fig. 70B). Dark band on narrowest part of hind femora not as pronounced as in *C. parviceps*. Fastigium deeply grooved (in *C. samalayucae* the fastigium is not deeply grooved) (Fig. 69C). Body length to end of forewings 16–22 mm in males, 22–29 mm in females.

HABITAT. Beaches, sandy tracts, and sand dunes. The type series was collected near San José Beach 40 miles SW of Ciudad Obregón, Sonora. Other specimens have been found on old sand dunes sparsely covered with vegetation at two sites in Baja California— 46 km W of Constitución and back of the beach near Mulege.

LIFE CYCLE. In lower Baja California, adults have been collected in May, July, August, and November. In Sonora adults were collected in May.

REFERENCES. Vickery 1969a.

Cibolacris samalayucae **Tinkham**

DISTRIBUTION. West Texas and northern Mexico. Presently known only from the vicinity of El Paso, Texas, both east and south of the Mexican border.

RECOGNITION. Very similar to *C. crypticus,* but fastigium not deeply furrowed. Pronotum with low tubercles and without two prominent tubercles at the front of the anterior sulcus. Median carina on prozona barely indicated (distinct in *C. crypticus*). Body pale, sand-colored.

HABITAT. Mesquite-stabilized sand hummocks on the Samalayuca Dunes south of El Paso and dunes east of El Paso. Tinkham reported that in dunes near La Noria, Chihuahua, the species was quite localized and rare and inhabited a low depression in gentle undulating dunes covered with short dead *Salsola*.

LIFE CYCLE. Adults were collected on June 26 and September 21.

REFERENCES. Tinkham 1961.

Cibolacris weissmani **Otte, n. sp.**   Pl. 10

DISTRIBUTION. This species is known only from Baja California. Its known range extends from about Cabo San Quintin in Baja California Norte to La Paz in Baja California Sur.

RECOGNITION. This species is very similar to *C. crypticus*. It differs as follows: Body length to end of forewings 14–17 mm in males (19–22 mm in *crypticus*), 21–25 mm in females (22–29 mm in *crypticus*). Body very finely speckled; no dark spots are as large as the eyes (some dark spots as large as the eyes or larger in *crypticus*). Hind tibiae sometimes whitish to yellowish, usually pinkish distally (whitish to bluish in *crypticus*). See Appendix IV for holotype data.

SPECIMENS. The 104 known specimens of this species are in the California Academy of Sciences, University of Michigan Museum of Zoology, British Museum, United States National Museum, Paris Museum, and Academy of Natural Sciences of Philadelphia. The specimens were collected at the following places. Baja California Norte: 68 km S of Rosarito on Highway 1, July 25, 1977 (Weissman). 7 km N of Guerrero Negro, July 10, 1978 (Weissman and Lightfoot). 21 km S of San Quintin, June 7, 1925 (Keifer). Baja California Sur: 11.2 km along dirt road to Scammons Lagoon off Highway 1, July 11, 1978 (Weissman and Lightfoot). 1.8 km W Guerrero Negro on road to Estero de San José, July 11, 1978 (Weissman and Lightfoot). Magdalena Bay, May 30, 1925 (Keifer). La Paz, March 4, 1968 (Viault).

HABITAT. Sand dunes.

LIFE CYCLE. Adults have been collected from March to July.

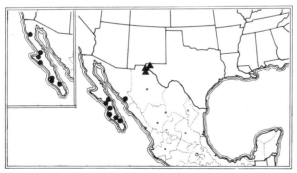

Cibolacris crypticus (circles), *C. samalayucae*
(triangles), and *C. weissmani* (inset)

Genus XERACRIS Caudell

DISTRIBUTION. Mojave and Sonoran deserts.

RECOGNITION. Small, sand-colored, finely spotted species. Females without stridulatory teeth on hind femora (present in *Cibolacris* females). Forewings extending beyond hind femora. Hind femora distinctly banded on top. Hind tibiae ivory or pale yellowish. Posterior margin of pronotal disk rounded. Lateral pronotal carinae absent, disk rounding gradually onto the lateral lobes. Lateral region of disk cut by three sulci. Median carina very low and usually cut by at least three sulci (Fig. 69D). Hind knees not dark.

REFERENCES. Caudell 1915, Jago 1971.

● **Identification of Xeracris Species**

minimus
1. Medial area of hind femora banded
2. Upper marginal area of hind femora with two dark bands (excluding knees)
3. Front half of lateral lobes with broad dark area.
4. Spurs on hind tibiae as in Fig. 70E
5. Posterior margin of pronotal disk slightly angulate (Fig. 69D)

snowi
1. Medial area of hind femora unbanded
2. Upper marginal area of hind femora with three or four dark bands (excluding knees)
3. Front half of lateral lobes spotted but without broad dark area
4. Spurs on hind tibiae as in Fig. 70D
5. Posterior margin of pronotal disk rounded (Fig. 69E)

Xeracris minimus (Scudder)   Pl. 10

DISTRIBUTION. Deserts of southern California, southern Nevada, and southwestern Arizona.

RECOGNITION. Sandy colored. Sexes similar. Body length to end of forewings 11–16 mm in males, 16–23 mm in females. Top of hind femora with two distinct dark bands (often indistinct in yellowish specimens); these bands divide the femora into equal thirds; the large anterior band usually extends ventrally onto the medial area. Dorsal field of forewings with an irregular scattering of small brown spots. Lateral lobes often with a large dark area in upper front quarter and with the markings from the two sides sometimes connecting across disk by a narrow dark band along the front side of the posterior sulcus. Metazona often white, especially on the lateral lobes. Hind tibial spurs as in Fig. 70E.

HABITAT. Desert and desert scrub; in southern California and Nevada usually inhabiting creosote desert. Lives on ground during cooler hours of the day and perches on vegetation to avoid extreme temperatures.

LIFE CYCLE. Adults from July to September.

REFERENCES. Scudder 1900b, Caudell 1915, Ball et al. 1942.

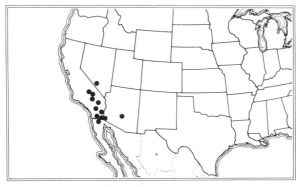

Xeracris minimus

Xeracris snowi **(Caudell)**   Pl. 10

DISTRIBUTION. Deserts of southern California, southwestern Utah, and western Arizona.

RECOGNITION. Pale, sandy colored species. Sexes similar. More speckled than *X. minimus*. Forewings with rows of small spots between veins. Top of hind femora (including knee) with at least three and often five dark bands. Unlike *X. minimus,* no dark bands extend onto outer medial area, which is usually nearly white. Lateral lobes highly speckled. Metazona not white as in *X. minimus*. Disk of pronotum usually with two dark marks between two posterior transverse sulci and with two very pale spots on the metazona adjacent to and behind the dark spots. Tibial spurs as in Fig. 70D. Hind tibiae pale yellow to white.

HABITAT. According to Ball et al. (1942), *X. snowi* is a rare species which has been found associated with the low-growing plant *Coldenia palmeri* in sandy areas of western Arizona.

LIFE CYCLE. Adults from July to September.

REFERENCES. Caudell 1915, Ball et al. 1942, Jago 1971.

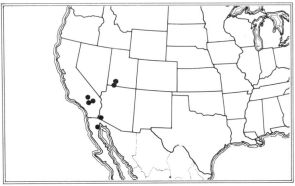

Xeracris snowi

Genus LIGUROTETTIX McNeill

DISTRIBUTION. Sonoran and Chihuahuan deserts.

RECOGNITION. Gray brown insects living on bushes. Anterior margin of forewings in male with larger oblique cells (Fig. 47). Hind femora banded on upper and inner faces with three black bands (including knees), and sometimes faintly banded on outer face. Lateral foveolae visible from above. Antennae filiform. Forewings extend well beyond the end of the abdomen. Lateral pronotal carinae absent, disk of pronotum rounds gradually onto the lateral lobes. Edge of pronotal disk cut by three sulci. Forewings almost unicolorous gray brown, sometimes very faintly marked with darker spots. Posterior margin of pronotal disk rounded.

REFERENCES. Rehn 1923, Jago 1971.

● **Identification of Ligurotettix Species**

coquilletti (Sonoran Desert)
1. Male forewings with very large cells near costal margin (Fig. 44A)
2. Body length to end of forewings 16–23 mm in males, 21–30 mm in females

planum (Chihuahuan Desert)

1. Male forewings with smaller cells near costal margin (Fig. 44B)
2. Body length to end of forewings 23–30 mm in males, 26–33 mm in females

---

Ligurotettix coquilletti **McNeill**   Pl. 10

TAXONOMY. Rehn 1923 recognized three subspecies of *L. coquilletti: L. c. coquilletti* McNeill, mainly from California; *L. c. cantator* Rehn, mainly from Nevada; and *L. c. kunzei* Caudell, mainly from Arizona and extreme southern California. It is difficult to separate the subspecies, and interested readers should consult Rehn's paper on this group. An undescribed species may exist on the Baja California peninsula. Two males I captured on creosote bushes produced a song quite different from that of typical *L. coquilletti* in Arizona and California. Each burst of stridulation consisted of ten or more rapidly delivered pulses. The typical *L. coquilletti* song consists of one to three pulses, more widely spaced (Otte 1970).

DISTRIBUTION. Sonoran Desert of Nevada, California, Arizona, Baja California, and Sonora.

RECOGNITION. Very similar to *L. planum,* but leading edge of forewings with very large cells in males (Fig. 44A). In Nevada and western Arizona two color forms coexist in the same bushes—one is mostly gray with mostly gray lateral lobes; the other has the dorsum very pale and the sides of the body nearly black; lateral lobes nearly black in the anterior or upper anterior part and ivory in the lower and posterior part. Body length to end of forewings 16–23 mm in males, 21–30 mm in females.

HABITAT. This small gray grasshopper inhabits the southwestern deserts, where it is usually found on creosote bushes (*Larrea divaricata*). It is sometimes also common on *Atriplex* shrubs and occasionally inhabits *Franseria* bushes, mesquite trees (*Prosopis*), and ironwood (*Olneya*).

BEHAVIOR. If population density is low, males of *L. coquilletti* are territorial. A bush rarely contains more than one male, and males fight for possession of the bush. Mutual repulsion is achieved through continuous stridulation, consisting of one, two, or three sharp clicking or zipping sounds (*zip . . . zip-zip-zip . . . zip-zip*). Males stridulate from shortly after sunup until about midnight. For details on the biology of this species see Otte and Joern (1975, 1977).

REFERENCES. Rehn 1923, Otte and Joern 1975, 1977.

Ligurotettix planum (Bruner)   Pl. 10

DISTRIBUTION. Chihuahuan Desert from southwestern New Mexico to states of Durango and Zacatecas, Mexico.

RECOGNITION. Very similar to _L. coquilletti,_ but cells along leading edge of forewings are not so large (Fig. 47B). Since these two species have largely nonoverlapping ranges and different habitats, there is little danger of confusing them. Body length to end of forewings 23–30 mm in males, 26–33 mm in females.

HABITAT. In west Texas, southern New Mexico, and extreme southeastern Arizona, _L. planum_ is commonly found on the bush _Flourensia cernua._ In Coahuila, Mexico, the species was common on three shrubs—_Flourensia cernua, Cordia parvifolia,_ and _Sericodes greggii_ (Otte and Joern 1977). It has never been found on creosote bushes, even though this shrub is very abundant in its habitat.

BEHAVIOR. Like _L. coquilletti, L. planum_ is territorial, with males defending bushes under lower population densities. Under high densities males are less aggressive, and a single bush may be inhabited by as many as four or five males, all but one of which is usually silent.

LIFE CYCLE. Adults from late summer to fall.

REFERENCES. Bruner 1904, Rehn 1923, Tinkham 1948, Otte and Joern 1975, 1977.

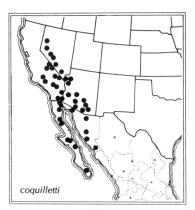 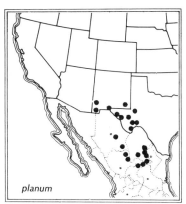

Ligurotettix coquilletti and _L. planum_

Amblytropidia Genus Group

The genera Boopedon, *Amblytropidia, Leurohippus,* and *Syrbula* are tentatively grouped together only because of their grouping in the first set of key couplets and because each genus cannot easily be grouped with other genera. The closer resemblance between *Amblytropidia* and *Leurohippus* may, however, suggest a more recent common ancestry.

Genus BOOPEDON Thomas

TAXONOMY. The genus includes eight species. The four species possessing a concave fastigium (*flaviventris, diabolicum, rufipes,* and *dampfi*) formerly belonged to *Morseiella,* while those with a largely convex fastigium belonged to *Boopedon.* Several groups of species are evident. *B. rufipes* and *B. dampfi* appear to share a recent common ancestry, as do *B. auriventris, B. gracile,* and *B. nubilum.* Females of the former two species share unique color patterns, and females of the latter three may be confusingly similar. *B. empelios* seems to be somewhat intermediate between *B. diabolicum* and *B. nubilum. B. gracile* and *B. auriventris* are also similar in coloration and differ mainly in wing length.

DISTRIBUTION. Central and southwestern United States to southern Mexico.

RECOGNITION. Disk of pronotum never possessing FDI. Top of head broadly rounded in lateral profile. Fastigium convex, flat, or slightly concave. Lateral foveolae poorly developed or obsolete, but foveolar area visible from above. Male forewings never extending beyond ends of hind femora and usually not reaching end of abdomen (exceptions are *B. gracile* and occasional males of *B. nubilum*). Female forewings rarely extending beyond middle of abdomen and usually shorter than head plus pronotum. Body length to end of femora more than 20 mm in males and more than 25 mm in females (except *B. rufipes,* which is considerably smaller). Longest apical spur on hind tibiae about twice as long as the other two. Lateral pronotal carinae diverge slightly on the metazona. Junction between pronotal disk and lateral lobes usually well marked, but the two may round into one another without producing distinct lateral carinae. The lateral carinal region cut by one, two, or three sulci.

REFERENCES. Jago 1971.

■ Key to Boopedon Species

MALES

1. Body color mostly black. Hind femora not banded on outer or upper face: 2

   Body not mostly black; if dark, then outer face of hind femora banded: 4
2. Hind tibiae red in distal half. Face not tan or brown: 3

   Hind tibiae black. Face tan or light brown: **diabolicum**
3. Lower side of hind femora reddish. Venter of abdomen brown to light brown: **empelios**

   Lower side of hind femora black. Venter of abdomen black: **nubilum**
4. Hind tibiae red in distal two-thirds: 5

   Hind tibiae black in distal two-thirds: **dampfi**
5. Legs yellow. Side of abdomen yellow: **flaviventris**

   Legs not yellow. Side of abdomen not yellow: 6
6. Lateral lobes with vertical light and dark bands. Abdomen often with reddish or pinkish coloration: 7

   Lateral lobes with horizontal light and dark bands. Abdomen without reddish or pinkish coloration: **rifipes**
7. Outer face of hind femora banded. Forewings reaching beyond middle of abdomen: **gracile**

   Outer face of hind femora not banded. Forewings not reaching beyond middle of abdomen: **auriventris**

FEMALES

1. Hind tibiae reddish: 2

   Hind tibiae not reddish: 6
2. Outer face of hind femora with dark bands: 3

   Outer face of hind femora without distinct dark bands: 4
3. Legs and side of abdomen yellow (southern Arizona and western Mexico): **flaviventris**

   Legs brownish, straw-colored, or blackish, side of abdomen without strong yellow coloration (northeastern Mexico to Oklahoma and Kansas): **gracile**
4. Upper half of lateral lobes mostly black, lower half yellowish. Lower marginal area of hind femora reddish (southern Mexico): **rufipes**

   Lateral lobes not as above. Lower marginal area of hind femora not red (northern Mexico and United States): 5

5. Forewings longer than head plus pronotum or reaching end of abdomen and overlapping medially. Lateral lobes without large, roughly triangular, dark marking. Body sometimes black:
                                                            nubilum
   Forewings shorter than head plus pronotum. Lateral lobes with large triangular dark marking (gray to blackish). Body color never black: **auriventris**
6. Lower half of hind femora reddish. Base of hind tibiae without yellow band: **diabolicum**
   Lower half of hind femora yellow. Base of hind tibiae with yellow band: **dampfi**

Boopedon nubilum (Say)   Pl. 11

DISTRIBUTION. Widespread through the short-grass prairies.

RECOGNITION. Sexual differences very pronounced. Males: Body shiny black. Face dark brown to black (pale in *B. diabolicum*). Hind femora not banded on medial area. Narrowest part of hind femora either with a pale ring or entirely black. Hind tibiae may be entirely black; black and red; or cream, black, and red. Lower marginal area of hind femora black (red in *B. diabolicum*). Forewing length variable, usually not longer than head plus pronotum and not reaching end of abdomen; rarely extending just beyond end of abdomen, but never extending beyond hind femora. Pronotal disk without clearly raised lateral carinae. Edge of disk cut by two or three sulci. Fastigium barely indicated, usually mostly convex, sometimes flat, and usually with a small median carina at the front (concave and without a median carina in *B. empelios*). Posterior margin of pronotum slightly angulate. Body length to end of femora 24–34 mm.

Females: Body usually pale brown or straw brown, but often partly green on the head, pronotum, and femora. Occasional females are dark brown or black. Forewing length variable, often shorter than head plus pronotum, never reaching end of abdomen. Forewings often touching medially and sometimes overlapping. Lateral lobes with black sulci. Crescents of hind knees black. Body length to end of femora 33–52 mm.

HABITAT. Short-grass prairies. Ball et al. (1942) report: "This is one of the most common and conspicuous of Arizona grassland species, probably the most important range grasshopper in southern

Arizona. The black males are active and conspicuous, while the larger females, unable to fly, remain in the grass and are not readily seen. [It is] more common in areas of *Andropogon, Aristida,* and other tall grasses than in areas of curly mesquite (*Hilaria*) and *Bouteloua.* Also found in cultivated areas on corn, sorghum, and wheat, and grass weeds such as Johnson grass (*Sorghum halepense*) and *Echinochloa* [and] common from 2,200 to 5,500 feet."

LIFE CYCLE. Adults from August to October in Texas, Arizona, and northern Mexico; from July to September in North Dakota, Nebraska, and Montana.

REFERENCES. Say 1825a, Thomas 1871a, Ball et al. 1942, Tinkham 1948, Brooks 1958.

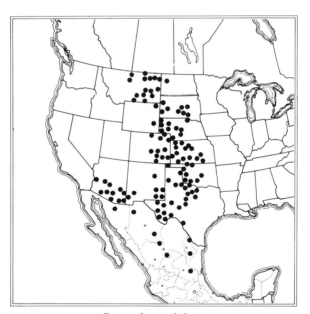

Boopedon nubilum

***Boopedon diabolicum* (Bruner)**   Pl. 11

DISTRIBUTION. Southwestern Mexico from Nayarit to Mexico City.

RECOGNITION. Males: Body color black. Face pale brown or cream (black in *B. nubilum*). Hind femora bright red on the lower marginal area. Posterior margin of pronotal disk straight (strongly convex in *B. nubilum,* convex or slightly angulate in *B. empelios*). Junction between pronotal disk and lateral lobes sharp (rounded in *B. nubilum* and *B. empelios*). Hind tibiae entirely black (red in distal half in *B. empelios*). Body length to end of femora 20–28 mm.

Females: Yellowish, brown, and black. Dorsum of body brownish, venter and lower sides yellowish. Head black behind the eyes. Lateral lobes with a large, roughly triangular black area in the upper two-thirds and with a narrow yellowish band along the front margin. Hind tibiae usually yellowish. Forewings oval, shorter than pronotum, and not overlapping medially. Body length to end of femora 28–37 mm.

HABITAT. Grasslands.

LIFE CYCLE. All adults in the Philadelphia Academy collection were collected in August and September.

REFERENCES. Bruner 1904, Hebard 1932.

Boopedon diabolicum

Boopedon empelios **Otte**   Pl. 11

DISTRIBUTION. Known only from three males collected 16 miles E of Navajoa, Sonora, on the highway to Alamos, August 20, 1964.

RECOGNITION. Based only on males; females unknown. Body mostly black, but venter of thorax and abdomen yellowish (unlike *B. nubilum* or *B. diabolicum*). Face black (pale in *B. diabolicum*). Hind tibiae black in proximal half, bright red in distal half, and without a pale ring near the proximal end. Hind femora yellow to orange along lower marginal area and lower side of the medial area (red in *B. nubilum*). Fastigium of vertex concave and with a definite transverse groove (largely convex in *B. nubilum*). Body length to end of femora 24–30 mm.

REFERENCES. Otte 1979b.

Boopedon flaviventris (Bruner)   Pl. 11

DISTRIBUTION. Southern Arizona through western Mexico to Michoacán.

RECOGNITION. Sexes similar in color. Body pale brown on dorsum, yellow on sides and venter. Hind femora strongly banded with black on the outer and inner face. Hind knees black. Hind tibiae black at base, then with a yellowish band, remaining three-quarters red or orange. Lateral lobes with a roughly triangular dark area on the upper side. Posterior margin of pronotum straight. Disk and lateral lobes do not meet at a sharp angle. Hind margin of male forewings blunt. Forewings of females oval, usually no longer than the pronotum, sometimes slightly overlapping and without stripes. Dorsum of female abdomen with two rows of black markings; each black mark contains an oblique ivory streak. Female pronotal disk often with two pale stripes, one along each margin. Body length to end of femora 20–29 mm in males, 27–44 mm in females.

HABITAT. This species reaches high densities in the grasslands of southern Arizona and is occasionally destructive to rangeland. Ball et al. (1942) note that it is "especially common in *Aristida,* grama grass (*Bouteloua* sp.) and *Andropogon scoparius* in swales in the desert grassland, but is sometimes found in dense growths of sacaton grass (*Sporobolus wrightii*) or Bermuda grass (*Cynodon dactylon*). At Arivaca one infestation was found to average two adults per square foot." Generally found in medium-length grasses and in grassy and lightly wooded areas.

REFERENCES. Bruner 1904, Hebard 1932.

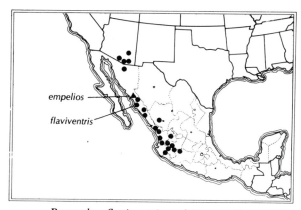

Boopedon flaviventris and *B. empelios*

Boopedon gracile **Rehn**   Pl. 11

DISTRIBUTION. Northeastern Mexico through central Texas to Oklahoma and Kansas.

RECOGNITION. Males: Forewings usually extend to end of abdomen. Hind femora with black bands on outer, upper, and inner faces, and with dark bands broader than pale bands. Hind knees black. Lateral lobes with a black triangular marking on the prozona from the top to the bottom of the lobe; a pale ivory band runs along the front margin. Hind tibiae black at base, then cream, then black, then reddish. Junction between pronotal disk and lateral lobes sharp and slightly raised into a ridge. Pronotal disk usually with pale lines along the lateral carinae (absent in *B. auriventris*). Body length to end of femora 24.0–37.0 mm.

Females: Most easily confused with *B. nubilum* females. Dorsum of forewings with pale longitudinal stripes (as in *B. nubilum*), but lateral field with dark spots (*B. nubilum* has pale spots). Body length to end of femora 30.0–48.0 mm. Body color much lighter than males— usually pale yellow brown or greenish and strongly marked with dark bands, stripes, and spots. Lateral lobes dark on the prozona between the first two vertical sulci and with a pale band along the front margin. Median pronotal carina usually dark. Pronotal disk usually with pale lines along lateral carinae. Medial area of hind femora with three broad dark bands. Upper marginal area largely unbanded. Hind knees black only on inner and outer crescents.

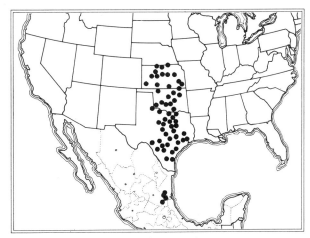

Boopedon gracile

Forewings pointed, overlapping medially, and variable in length
—usually shorter than head plus pronotum, but very occasionally
reaching end of abdomen.

HABITAT. Short- to medium-length grasslands.

LIFE CYCLE. Adults in July and August.

REFERENCES. Rehn 1904.

***Boopedon auriventris* McNeill**   Pl. 11

DISTRIBUTION. Iowa, eastern Nebraska and Kansas, Missouri,
Oklahoma, north central Texas, and western Arkansas.

RECOGNITION. Males and females similarly colored: Male forewings usually shorter than head plus pronotum, but sometimes
slightly longer. Lateral lobes with a strong triangular dark area on
most of the surface; the triangle extends into the metazona (unlike
B. gracile) and is bordered front and back by ivory bands. Hind margin of lateral lobes gray brown. Upper and outer faces of hind femora slightly banded, inner face with two black bands. Hind knees
entirely black in males, black on outer crescents in females. Female
forewings slightly pointed, sometimes barely overlapping medially
and, except for a few small dark spots, uniformly gray brown. Sides
of abdomen in males reddish; in females, with large irregular black
markings. Lateral pronotal carinae of males never pale. Male hind
tibiae black, cream, and black in proximal half and reddish in distal
half. Body length to end of hind femora 25–32 mm in males, 35–
42 mm in females.

HABITAT. Grasslands and grassy clearings in open woods from

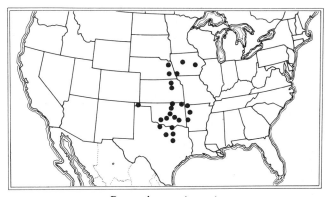

Boopedon auriventris

Iowa to Texas. Isely (1937) found it in the blackland prairie belt and eastern cross timbers of Texas, and Morse (1907) wrote: "This is an interesting sylvan species peculiar to the southern half of the forested region between the Mississippi River and the Great Plains. It was first met with at Mena (Arkansas), on the eastern slope of the foothills of Rich Mountain, in an open forest of pine and deciduous trees [and] at Hailyville and Caddo Hill, Oklahoma, in similar rocky, deciduous woodlands. At Magazine Mountain (Arkansas) it was one of the notable assemblage of species which had gathered in the shrubby growth at the edge of the summit cliff. It is of sluggish disposition, and but few are seen, but when aroused it leaps powerfully, often two or three times in succession. The males appear ridiculously small beside their huge mates."

LIFE CYCLE. Adult season is from July into August.
REFERENCES. McNeill 1899, Morse 1904, Isely 1937.

Boopedon dampfi (Hebard)   Pl. 11

DISTRIBUTION. Extreme southern Mexico.

RECOGNITION. Yellow and black species, similar to *B. rufipes* but with black hind tibia. Top of body with two narrow yellow stripes running from the backs of the eyes along the lateral margins of the pronotal disk and onto the wings. Sides of body with a broad black band (with irregular lower margin) extending from behind the eyes onto the forewings. Hind femora mostly yellowish, with black knees and two poorly defined bands along the upper part. Hind tibiae black or dark brown except for a yellow band near the base (orange or red in *B. rufipes*). Forewings in males as long as head plus pronotum or slightly shorter. Forewings in females with rounded posterior mar-

Boopedon dampfi

gin, shorter than pronotum, and slightly overlapping medially. Body
length to end of femora 23–33 mm in males, 30–39 mm in females.
REFERENCES. Hebard 1932, Jago 1971.

Boopedon rufipes (Hebard)   Pl. 11

DISTRIBUTION. Mexico, from Morelos to Oaxaca.

RECOGNITION. Body color brownish on dorsum, yellow on
venter. Hind margin of pronotal disk straight or slightly concave.
Sides of body with a broad black band from the eyes onto the forewings. Hind tibiae orange-red, black at the base. Hind femora orange to red along lower marginal area and with black knees. Lateral
carinae usually cut by two sulci. Lateral margins of pronotal disk
sometimes with a pale line. Male forewings about as long as head
plus pronotum or shorter and truncated at the end. Forewings of females shorter than pronotum, somewhat oval, and slightly overlapping medially. Fastigium concave. Body length to end of femora 17–
24 mm in males, 24–26 mm in females.

HABITAT. No published information. Probably open grassy
woodlands.

LIFE CYCLE. All specimens in the Philadelphia Academy collection were collected in August and September.

REFERENCES. Hebard 1932.

Boopedon rufipes

Genus AMBLYTROPIDIA Stål

TAXONOMY. This relatively large genus of twenty-three named
species is best represented in South America. When the genus is re-
vised, the number of valid species will probably be reduced to not
more than fifteen. Only three species are known from north of the
Panama Canal, although eleven were described from the region.

DISTRIBUTION. Southern third of the United States to Argentina.

RECOGNITION. North American species only. Body color brownish, usually several shades of brown, but sometimes with blackish, dark olive green, and reddish areas. Body length to end of forewings 19–27 mm in males, 27–38 mm in females. Antennae filiform. Front of head without lateral foveolae. Region normally occupied by the foveolae hidden from above by the lateral margins of the fastigium. Fastigium with a low median carina, often extending back to the pronotum. Fastigium either flat or convex and with lateral fastigial carinae distinct (*A. trinitatis, A. elongata*) or indistinct (some males and all females of *A. mysteca*). Frontal costa broad, convex, and gradually rounding onto top of head. Top of head often with three closely parallel low carinae; the lateral two may be poorly defined, and the three ridges together may sometimes be joined into a single broad band raised slightly above the rest of the head. Side and front of the head uniformly brown. Sides of pronotal disk nearly parallel, with very slight posterior divergence. Lateral carinae may be weakly or moderately developed, but the disk always meets the lateral lobes at a sharp angle. Lateral carinae cut by two or three sulci. Posterior margin of disk angulate. Forewings always extending to end of the abdomen or beyond. Hind femora without crossbands, but knees sometimes black. Hind femora very smooth and polished in appearance.

REFERENCES. Saussure 1861, Stål 1873, Hebard 1923, 1926, 1932, 1933.

- **Identification of Amblytropidia Species**

mysteca (U. S. to Costa Rica)
  1. Carinae of fastigium indistinct in both sexes, less distinct in females
  2. Hind knees brown
  3. Middle antennal segments short; length less than one and a half times width
  4. Lateral pronotal carinae cut by two sulci
trinitatis (southern Mexico to Colombia)
  1. Carinae of fastigium distinct in both sexes
  2. Hind knees entirely black
  3. Middle antennal segments long; length is two and a half to three times width
  4. Lateral pronotal carinae cut by three sulci
elongata (central Mexico)
  1. Carinae of fastigium distinct in both sexes
  2. Hind knees black on sides

3. Middle antennal segments intermediate in length; about twice as long as width
4. Lateral pronotal carinae cut by two sulci

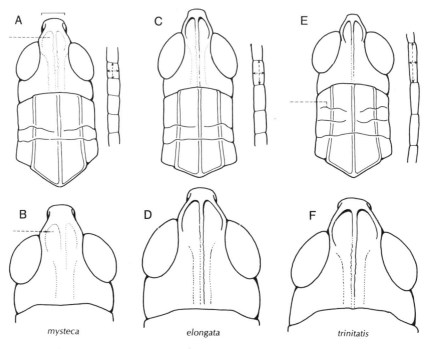

mysteca   elongata   trinitatis

Fig. 71. *Amblytropidia* species. Comparison of head, pronotal, and antennal characters.

Amblytropidia mysteca (**Saussure**)   Pl. 12

DISTRIBUTION. Ranges across the southern United States from Arizona to Virginia and through most of Mexico and parts of Central America.

RECOGNITION. Fastigium convex with lateral and median carinae often barely indicated, especially in females. In *A. elongata* and *A. trinitatis* the fastigium has distinct ridges. Antennae shorter than head plus pronotum. Length of middle antennal segments less than one and a half times their width. Lateral carinae cut by two sulci. Hind femora often very dark on outer medial area and tan on upper marginal area. Hind tibiae lighter brown in basal two-thirds and turning black distally.

The following color morphs are common: (a) pale brown on top of body and dark brown or black on the sides; (b) gray on top and black on the sides; (c) entire body uniformly brown; (d) black on top and brown on the sides. Body length to end of wings 20–27 mm in males, 27–36 mm in females.

HABITAT. Blatchley (1920) wrote that in the southeastern United States, "Adults often fly long distances and frequently dive headlong into a tuft of grass where they attempt to burrow out of sight. In the pine woods . . . they often remain motionless until closely approached, depending upon their color resemblance to that of the pine needles for protection." In Virginia, Fox (1914) found the species in late autumn "in low grassy undergrowth of open woodlands of oak, loblolly pine and sweet gum." In Texas I have found it in thickly grassed areas in the juniper-oak savanna and usually in dense grasses not more than knee high.

LIFE CYCLE. In Florida the species overwinters in the late nymphal and adult stages, and the egg stage occurs in the summer months. In Texas adults may be found during most months, but in central Texas they are abundant in the early spring. In Virginia adults have been taken in the late fall and spring months.

REFERENCES. Saussure 1861, Rehn 1902a, Blatchley 1920.

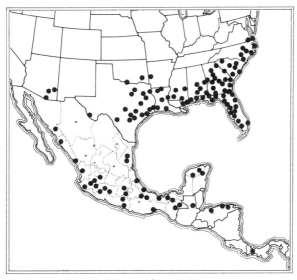

Amblytropidia mysteca

Amblytropidia trinitatis **Bruner**   Pl. 12

DISTRIBUTION. Veracruz and Tabasco, Mexico, south to Guyana, Trinidad, Venezuela, and Colombia.

RECOGNITION. Fastigium with well-developed medial and lateral carinae in both sexes. Lateral pronotal carinae cut by three sulci. Antennae longer than head plus pronotum in both sexes and in males about one and a half times as long. Length of middle antennal segments at least three times the width. Males dark olive green to blackish on the sides of head and thorax. Hind knees black, medial area orange or red, and upper marginal area usually much lighter than medial area. Hind tibiae brown, becoming blackish in distal half. Females more variable in color than males; some are unicolorous brown, others are nearly black on the sides and brown on top, and a few are darker on top of the body than on the sides. Body length to end of forewings 19–27 mm in males, 27–38 mm in females.

HABITAT. Probably mainly a forest-edge species. Labels on specimens collected in Colombia indicate a forest habitat.

LIFE CYCLE. Adults may be found throughout the year.

REFERENCES. Bruner 1904, Hebard 1923, 1932.

Amblytropidia trinitatis

Amblytropidia elongata **Bruner**   Pl. 12

DISTRIBUTION. Central Mexico.

RECOGNITION. Differs from *A. mysteca* as follows: Fastigium of vertex with clearly defined anterior and medial ridges (Fig. 71); disk of pronotum relatively longer. Middle antennal segments in males about twice as long as wide. Hind knees of males and a few females black on the sides. Differs from *A. trinitatis* in having two instead of three sulci cutting the lateral carinae and having relatively shorter antennal segments. Body length to end of forewings 20–29 mm in males, 29–37 mm in females.

LIFE CYCLE. Adults in the Philadelphia Academy collection were taken between July and September.

REFERENCES. Bruner 1904.

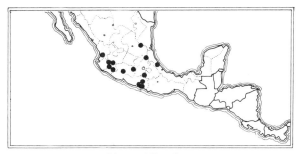

Amblytropidia elongata

Genus LEUROHIPPUS Uvarov

TAXONOMY. This genus, which is similar and probably related to *Amblytropidia,* is restricted to South America and the Caribbean region. *L. stoneri* (formerly *Caribacris*) is the only known Caribbean species. It is now confined to relatively undisturbed hillsides on Antigua.

DISTRIBUTION. South America and Caribbean region.

Leurohippus stoneri (Caudell)   Pl. 12

DISTRIBUTION. Known only from the southeastern hills of Antigua. These hills are part of a historical monument and are protected from goats and cattle.

RECOGNITION. Forewings not reaching end of abdomen but longer than head plus pronotum. Body pale brown, but dark brown at the upper ends of the lateral lobes. Antennae ensiform. Lateral foveolae small, barely distinguishable, and invisible from above. Fastigium mostly convex, but with strong ridges around front and sides and with a well-developed median ridge. Disk of pronotum with parallel sides and with well-developed lateral carinae cut by two sulci. Posterior margin of disk angulate. Disk pale brown, much lighter than the lateral lobes. Hind femora and tibiae light brown and unbanded.

REFERENCES. Caudell 1922, Jago 1971.

Genus SYRBULA Stål

DISTRIBUTION. Eastern, southern, and southwestern United States and most of Mexico.

RECOGNITION. Forewings always extending beyond end of abdomen (except in some egg-laden females), but rarely extending beyond hind femora. Hind tibiae with sixteen to twenty-two outer spines. Fastigium of vertex largely convex and with a median carina. Lateral foveolae not visible from above. Lateral pronotal carinae well developed; slightly to strongly constricted on prozona (in *S. admirabilis* sometimes nearly parallel), and always cut by two or three sulci. Frontal costa slightly convex, flat, or slightly concave. Male antennae filiform, but slightly clubbed in males of *S. montezuma* and *S. admirabilis* and ensiform in *S. festina*. Female antennae slightly ensiform. Male subgenital plate distinctly pointed, but not elongated as in *Achurum*.

REFERENCES. Otte 1979b.

● **Identification of Syrbula Species**

MALES
admirabilis
    1. Hindwings transparent
    2. Narrowest part of hind femora with pale band
    3. Side of head with white streak behind eye
    4. Lateral lobes white along front margin and with white spur.
montezuma
    1. Hindwings black
    2. Narrowest part of hind femora without pale band
    3. Side of head black or black and green behind eye
    4. Bottom of lateral lobes with a narrow ivory band; top third green or blackish
festina
    1. Hindwings transparent
    2. Narrowest part of hind femora without pale band
    3. Side of head with fine black and white lines behind eye
    4. Lateral lobes green with narrow horizontal black line on prozona

FEMALES

admirabilis

1. FDI extending from front to back margin
2. Lateral carinae slightly constricted (Fig. 72E)
3. Lateral field of forewings with spots or wavy dark markings
4. Hindwings transparent

montezuma

1. FDI forming triangles in posterior half of disk
2. Lateral carinae strongly constricted (Fig. 72F)
3. Lateral field of forewings with spots or wavy dark markings
4. Hindwings black

festina

1. FDI extending from front to back margins, narrow in central part
2. Lateral carinae moderately constricted (Fig. 72G)
3. Lateral field of forewings without dark markings
4. Hindwings clear

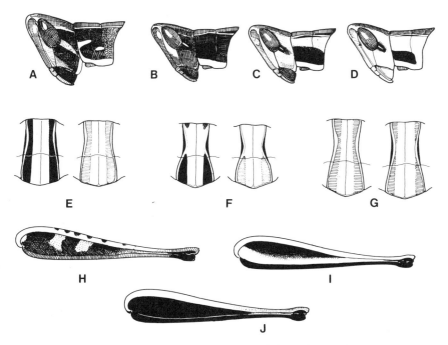

Fig. 72. *Syrbula* species. A, *S. admirabilis* male; B, C, D, *S. montezuma* male color variation; E, *S. admirabilis* females; F, *S. montezuma* females; G, *S. festina* females; H, I, J, *S. montezuma* male hind femur variation.

Syrbula admirabilis (Uhler)   Pl. 12

DISTRIBUTION. Eastern United States and Mexico.

RECOGNITION. Males: Side of head with an ivory stripe from back of eye to lateral lobe. Lateral lobes with a pale band along bottom, curving up along front margin, and with a white or ivory spur extending from middle front margin toward middle. Usually a small white spot just below midpoint of lateral lobes (*S. montezuma* lacks white spot and spur). Lateral field of forewings usually with large, though often indistinct, spots. Hind femora with a pale band on narrowest part next to the knees. Lateral carinae slightly constricted on the prozona (Fig. 72A) (strongly constricted in *S. montezuma*). Forewings transparent (black in *S. montezuma*). Body length to end of femora 25–31 mm.

Females: Background color usually greenish, but some individuals are brown or tan. Lateral carinae of pronotum pale and bordered medially along its entire length with a black band (in *S. montezuma*, disk has triangular FDI). Top half of lateral field of forewings darkened by a row of large conjoined dark spots (similar to *S. montezuma*). Hindwings largely transparent (black in *S. montezuma*). Body length to end of femora 34–49 mm.

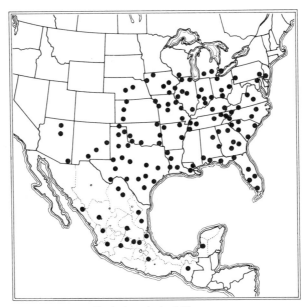

Syrbula admirabilis

HABITAT. Dry upland with poor soil and scanty vegetation (Rehn and Hebard 1916, Blatchley 1920). In Texas it is common in knee-high grasses in open areas.

BEHAVIOR. Males of *S. admirabilis* have very characteristic calling songs consisting of two-part units (*t-k* . . . *t-k* . . . *t-k*) repeated fifteen to sixty times. Each pair of sounds seems to be produced with a single stroke of the legs. Sexually receptive females either approach calling males or answer the male with their own softer stridulation. Courtship involves one of the most complex behavioral rituals known in insects (Otte 1972).

Males are much more likely to escape by flying than females, and after landing usually remain perched on vegetation. Egg-laden females usually prefer to hide in thick vegetation instead. Younger females fly readily, but usually dive into thick vegetation and creep along the ground after the first flight.

LIFE CYCLE. Egg-overwintering. Adults from midsummer to late fall in south-central states, and Texas.

REFERENCES. Bruner 1904, Rehn and Hebard 1916, Blatchley 1920, Otte 1970, 1972.

Syrbula montezuma (**Saussure**)   Pl. 12

DISTRIBUTION. Colorado, Arizona, New Mexico, Texas, and south to Oaxaca, Mexico.

RECOGNITION. Males: Blackish with pale brown or green markings. Two color morphs are common (Fig. 72B, C, D). In one form, face and side of head are mostly green with a black band behind eye, lateral lobes with a broad green band along upper side and an ivory band along lower side, separated by a black band. In the other form, side of head is largely dark, and lateral lobes mostly dark except for a ventral ivory band. Forewings largely black but with a white or pale green horizontal streak directly above hind coxa. Hindwings black. Hind femora without a pale band next to knee. Body length to end of femora 21–32 mm.

Females: Very similar to *S. admirabilis,* but differing as follows: hindwing black; disk of pronotum strongly constricted and darkened only on the metazona (in *S. admirabilis,* black bands run the entire length of the lateral carinae). Body length to end of femora 30–46 mm.

HABITAT. In Arizona *S. montezuma* inhabits taller grasses of desert grasslands. In central Texas it frequently inhabits grasses in small woodland openings. In west Texas and Mexico it is often found on rocky and grassy hillsides in grasslands and semi-deserts.

BEHAVIOR. Male song consists of a series of loud ticks and buzzes (approximately *tick-zzzz, tick-zzzz* . . . *tick, tick, tick, tick, tick*). In some songs the final tick series is omitted. Courtship, as in *S. admirabilis,* is extraordinarily complex (Otte 1972).

LIFE CYCLE. Egg-overwintering. Adults from midsummer to late fall.

REFERENCES. Bruner 1904, Tinkham 1948, Ball et al. 1942, Otte 1970, 1972, 1979b.

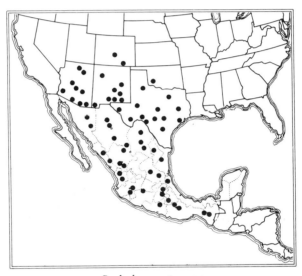

Syrbula montezuma

Syrbula festina **Otte**   Pl. 12

DISTRIBUTION. Known only from Oaxaca and Chiapas, Mexico.

RECOGNITION. Males: Body largely green. Lateral pronotal carinae moderately constricted (as in females, Fig. 72G). Head largely green, but top of head with a median ivory band bordered on either side by very thin dark lines. Antennae dark brown to black but with pale basal segments. Side of head green, but with an ivory streak joined to a black streak from back of eye toward lateral lobes. Lateral lobes green with a black horizontal line dividing prozona into equal parts. Disk of pronotum with a pale brown, ivory, or pale green median band bordered on either side by darker pigmentation. Forewings green, grayish or a combination; dorsal field of forewings darker than sides; sides with a pale green or white horizontal streak above base of hind legs. Hind femora unbanded, greenish in proximal half, reddish brown in distal half, with dark knees and without a pale band before the knees. Body length to end of femora 26–29 mm.

Females: Rather similar to *S. montezuma*, but hindwings not black; pronotal disk about as wide at front as at back (Fig. 72G); sides of forewings without spots. Body length to end of femora about 40 mm.

LIFE CYCLE. All adults in the Philadelphia Academy and University of Michigan collections were taken in August and September.

REFERENCES. Otte 1979b.

Syrbula festina

Mermiria Genus Group

Three genera are placed in this group: *Mermiria, Achurum,* and *Pseudopomala*. These genera all have the lateral foveolae hidden from view, are very slender, have strongly ensiform antennae, distinct and parallel lateral carinae, medium to broad postocular bands, and usually a median carinula on the fastigium. They live in medium to tall grasses.

Genus MERMIRIA Stål

TAXONOMY. Although the species recognized here can be distinguished without much difficulty, their relationships are probably more complicated than here depicted (Jago 1969).

DISTRIBUTION. Southern two-thirds of United States and northern Mexico.

BEHAVIOR. This genus inhabits medium to tall grasses or sedges. All species are good flyers and seem to rely more on active escape than on camouflage in avoiding predation. Once individuals have been flushed, they tend to keep flying well ahead of approaching collectors. Occasionally they hide behind grass stems or dive into the undergrowth.

RECOGNITION. These are among the largest New World Gomphocerinae, rivaled in size only by *Rhammatocerus*. Body length varies from 25–38 mm in males and from 32–57 mm in females. All species with a strong postocular dark band. Antennae ensiform. Lateral foveolae not visible from above. Fastigium of vertex usually shallowly grooved around the anterior margins and with a low, sometimes indistinct, median carina. Dorsum of body sometimes with a dark central stripe. Forewings always extending at least to the end of abdomen (except in egg-laden females). Hind femora usually extending slightly beyond end of abdomen. Hind tibiae reddish and bearing twenty or more outer spines. Hind femora not banded.

REFERENCES. Stål 1873, Scudder 1899b, Morse 1904, Rehn 1919, Jago 1969.

■ Key to Mermiria Species

EAST OF THE MISSISSIPPI RIVER

1. Lateral pronotal carinae distinctly raised and cut by two sulci. Postocular bands pink to red: **picta**
   Lateral pronotal carinae indistinct or absent, and edge of disk cut by three sulci. Postocular band dark brown to black: **2**
2. Dorsum of body with central dark stripe. Dorsal field of forewings more pale along lateral margins than medially:

                                                                    intertexta
   Dorsum of body without median dark band. Dorsal field of forewings dark along margins and with pale median band:

                                                                      bivittata

WEST OF THE MISSISSIPPI RIVER

1. Dorsum of body with two lateral ivory stripes which converge and join on forewing. Side of forewings with two white streaks, one above base of hind leg and another above abdomen: **texana**
   Dorsum of body without two distinct ivory bands. Side of forewings either with one white streak above base of hind leg or without white streaks: **2**

2. Disk of pronotum bordered by distinct lateral carinae cut by two
   sulci: **picta**
   Disk of pronotum rounding gradually onto lateral lobes, and edge
   of disk cut by three sulci: **bivittata**

Mermiria bivittata (**Serville**)   Pl. 13

TAXONOMY. The taxonomic status of *M. bivittata* is discussed in
detail by Jago (1969), who recognized two subspecies, *M. b. bivi-
tatta* and *M. b. malculipennis*. Jago described the distributional relations between the two subspecies, suggesting that the species
needs further study.

DISTRIBUTION. Widespread from California to the Carolinas and,
in the prairies, ranging north to Montana and North Dakota.

RECOGNITION. Coloration in this species is quite variable. West
of the Mississippi it may be distinguished from *M. texana* and *M.
picta* as follows: Dorsum mostly pale; this pale area becomes much
narrower and forms a median stripe on the forewings. Pale area on
dorsum sometimes with a median dark stripe, usually indistinct,
which does not extend onto forewings (dorsum in *M. texana* with
two lateral white bands and a median dark band). The lateral field of
forewings with one horizontal white streak or none (*M. texana* has
two white streaks; *M. picta* has none). *M. bivittata* differs from *M.
picta* as follows: Disk of the pronotum rounds off onto lateral lobe
(*M. picta* with distinctly raised carinae). Margin of pronotal disk always cut by three sulci (two in *M. picta*).

East of the Mississippi *M. bivittata* can be distinguished from *M.
intertexta* as follows: dorsum of the head and pronotum without a
wide dark median band, but sometimes with a thin dark line on the
median pronotal carina (*M. intertexta* with a strong median band
from front of head onto forewings). Dorsal field of forewing with a
narrow pale band which widens suddenly at pronotum. Body length
to end of forewings 28–38 mm in males and 39–56 mm in females.

HABITAT. According to Rehn (1919), the species inhabits areas of
"rich, high grass, with or without intermingled weeds, and from distinctly maritime (Florida) to relatively hilly or at least rolling en-
vironments (Oklahoma) and Texas to North Dakota" (Rehn 1919).
In Florida it inhabits *Juncus* along tidal inlets, post-oak groves and
forests, and grassy cover in open short-leaf pinewoods.

LIFE CYCLE. Adults from late June into October. In Kansas
nymphs are found from May 8 to June 20 and adults from July 1 to
August 3.

REFERENCES. Serville 1839, Bruner 1890, Rehn 1919, Jago 1969, Otte 1970.

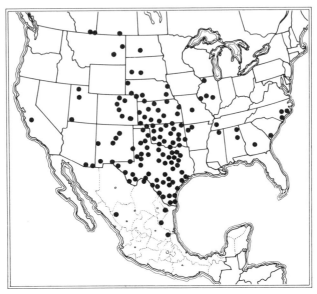

Mermiria bivittata

Mermiria picta **(Walker)**   Pl. 13

DISTRIBUTION. Widespread in central and southeastern United States and northern Mexico. Jago (1969) recognized two subspecies: *M. p. picta* (Walker) in the southeastern United States and *M. p. neomexicana* Thomas in the central and western states.

RECOGNITION. Side of forewing always without a white streak. Lateral pronotal carinae well developed and cut by two sulci. Post-ocular dark band bordered above and below by very thin pale lines. Body color in southeastern United States usually green with reddish dark bands, and dorsum usually with a well-developed median dark band. From Texas northward, body color is brownish; in this region *M. picta* may be confused with *M. bivittata* or *M. texana,* but it differs as follows: Side of wings without white streaks; lateral carinae well developed and cut by two sulci; top of body without two pale bands. Body length to end of forewings 30–41 mm in males, 46–57 mm in females.

HABITAT. In the southeastern United States *M. picta* inhabits tall

grasses in pine or mixed woods and occasionally the grasses along
borders of swampy areas. Rehn (1919: 85) wrote: "It also occurs in
associated gallberry and similar bushes in its preferred environment,
and persists in grasses and oak sprouts after the higher covering forest has been removed. Rarely it appears to invade grasses and
bushes growing up in oil fields, and rather infrequently is found in
sandy barrens of low oak and pine, where it occurs in the scant grass
and oak sprouts." Farther west it inhabits coarse grasses of the drier
prairie and is often locally abundant. It also inhabits hills and slopes,
even being considered by Bruner to be partial to hilltops. Hart reported it in Illinois on bunch grass (*Panicum virgatum*) in blowouts
between sand dunes.

LIFE CYCLE. Adults from July into November.

REFERENCES. Bruner 1897, Hart 1906, Rehn 1919, Jago 1969.

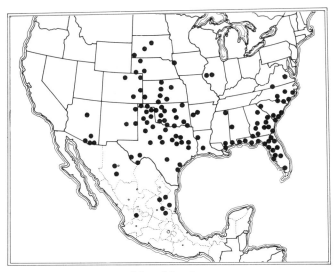

Mermiria picta

Mermiria intertexta **Scudder**   Pl. 13

DISTRIBUTION. Coastal region from New Jersey to Florida.

RECOGNITION. Dorsum with a median dark band which is most
pronounced on the head (also present in *M. picta,* but lacking in
eastern *M. bivittata*). Side of forewings with a single pale streak
above base of hind leg (lacking in *M. picta*). Lateral pronotal carinae

indistinct and usually cut by three sulci (distinct and cut by two sulci in *M. picta*). Postocular band brown to black (*M. picta* with reddish postocular band and a pale green background). Body length to end of forewings: 32–38 mm in males, 33–58 mm in females.

HABITAT. Coarse high grasses and reeds in marshy depressions and boggy conditions; also low oak, bayberry, palmetto, and briars among dunes (Rehn 1919). Overlaps geographically but little ecologically with *M. picta* and *M. bivittata*.

LIFE CYCLE. Adults from late July through September (New Jersey), early July to mid-November (Florida).

REFERENCES. Scudder 1899b, Rehn 1919, Jago 1969, Otte 1970.

Mermiria texana **Bruner**   Pl. 13

DISTRIBUTION. Southwestern United States and northern Mexico.

RECOGNITION. Top of body with two pale ivory bands which converge at the midsection of forewings. Lateral surface of forewings with two white streaks, one between the C and Sc veins directly above the base of the hind leg, the other above the R vein and above the abdomen. Top of hind femora with two dark spots on the upper carina. Body length to end of forewings 27–37 mm in males, 38–50 mm in females.

HABITAT. A hillside species in the arid southwest, *M. texana* inhabits stony and rocky slopes with a scattered cover of short grass

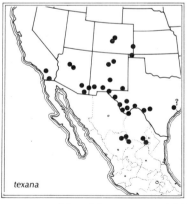

Mermiria texana and *M. intertexta*

or bunch grass and with other plants such as lecheguilla, agave, bear grass, ocotillo, sotol, and various cacti (Rehn 1919: 70). In Arizona it ranges from low-lying plains to 7,000 feet or more, at the higher elevations living in openings in the scrub-oak association.

REFERENCES. Rehn 1919, Jago 1969.

Genus ACHURUM Saussure

DISTRIBUTION. Florida to Arizona and Mexico.

RECOGNITION. Face very strongly slanting and undercut. Antennae strongly ensiform and, near the base, triangular in cross section. Head longer than pronotum. Front tibiae much shorter than top length of head. Abdomen extending to or beyond ends of hind femora. Male subgenital plate very long (except *A. sumichrasti*). Forewing length variable—very short in *A. carinatum* and very long in *A. sumichrasti*. Hind knees with pointed projections. All species display color polymorphisms, with green and pale brown morphs predominating. Males in the green phase are green on the dorsum and brown on the sides. Lateral carinae distinct and parallel. Differs from *Paropomala* in lacking a prosternal spine between the front legs, in having well-defined lateral carinae, and in having the front of the face concave in lateral profile.

REFERENCES. Jago 1969.

• Identification of Achurum Species

sumichrasti
1. Forewings extending beyond end of abdomen
2. Male subgenital plate shorter than front femora
3. Hind knees with long upper and lower lobes

minimipenne
1. Forewings not reaching end of abdomen, longer than head, overlapping dorsally
2. Male subgenital plate shorter than front femora
3. Hind knees with short upper and lower lobes

carinatum
1. Forewings shorter than head, not overlapping dorsally
2. Male subgenital plate longer than front femora
3. Hind knees with long upper lobes, short lower lobes

Achurum sumichrasti **(Saussure)**   Pl. 14

DISTRIBUTION. Southern regions of Texas, New Mexico, and Arizona, and most of Mexico.

RECOGNITION. Extremely slender, with long pointed forewings extending well beyond the abdomen. Body color usually green, brown, or a combination. Some individuals yellowish. Individuals may be almost uniformly pale green or pale brown, or pale on the dorsum and darker on the sides. Some have a pronounced dark stripe running along the upper sides from behind the eyes onto the forewings. Individuals from central and southern Mexico are often quite dark on the side of the body. Antennae strongly ensiform. Hind knees with upper and lower pointed projections. Body length to end of forewings 28–40 mm in males, 38–48 mm in females.

HABITAT. Tinkham (1948) gave the following account: "[In west Texas] this rare species was found in one particularly steep cut on the north slope of a high plateau at about a 5,500 foot elevation. Its habitat was limited to an altitudinal range of about 100 feet where it was found only in the tall grass in the bottom of the cut. [It] has a curious habit of making short sudden jumps from grass stem to grass stem and then 'freezing' or remaining rigid with its antennae aligned forward and pressed against the stem: their graceful form blending perfectly with their surroundings where they are almost impossible to locate unless disturbed." In southeastern Arizona the species may be "found clinging to stalks of coarse grasses on rocky slopes in des-

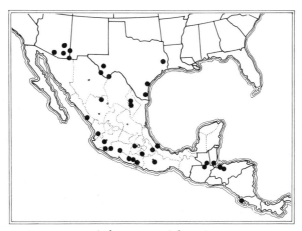

Achurum sumichrasti

ert grassland of the Lower and Upper Sonoran zones [where it is] commonly taken for *Andropogon* sp. *A. barbinodis, Muhlenbergia emersleyi, Eragrostis* sp. and *Sporobolus* sp'' (Ball et al. 1942).

LIFE CYCLE. In Arizona it overwinters in the nymphal stage, and adults are found in June and July.

REFERENCES. Ball et al. 1942, Tinkham 1948, Jago 1969.

Achurum carinatum (**Walker**)   Pl. 14

DISTRIBUTION. Florida. *A. c. carinatum* (Walker) ranges through northern Florida, and *A. c. brevipenne* (Thomas) through southern Florida. According to Jago (1969) the transition between the two subspecies is in the vicinity of Dixie, Hamilton, and Baker counties. He wrote: "The distribution of subsp. *brevipenne* is of interest. In the extreme south of Florida it shuns the everglade region and Cyprus Savanna . . . The Suwanee R. and St. Marys R. provide an almost complete ecological barrier in the north, their head-waters expanded to form the Okefenokee Swamp, so that this flightless species has given rise to two quite distinct subspecies separated by a physical and ecological gap."

RECOGNITION. Body extremely slender. Forewings minute, shorter than the head in *A. c. brevipenne* males and about as long as the head in *A. c. carinatum* males. Forewings not overlapping dorsally. Eyes set about one eye length forward of the base of the mandible. Antennae large and strongly ensiform. Subgenital plate longer than front femora. Hind knees with long dorsal projections and either short lower projections or none. Males either tan or pale green dorsally and darker brown on the sides. Females are highly variable in color. The following morphs are known: tan on top and green on the sides; entirely green; entirely tan; black on top and green on the sides. Tan females may be strongly spotted with black dots. Body length to end of abdomen 26–36 mm in males, 33–50 mm in females.

HABITAT. According to Rehn and Hebard (1916), *A. c. carinatum* is widely distributed throughout the undergrowth of the long-leaf pine forests and sometimes inhabits somewhat damp situations. It has also been found in short grass of waste land, in sandy tracts of scrub oaks, in mixed oak and pine woods, and in a sandy field of short grass. *A. c. brevipenne* inhabits low undergrowth in pine-woods.

LIFE CYCLE. In Georgia and northern Florida *A. carinatum* over-

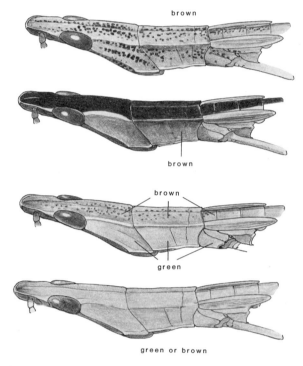

brown

brown

brown

green

green or brown

Fig. 73. Pattern variation in *Achurum carinatum* females.

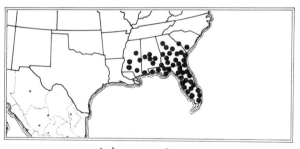

Achurum carinatum

winters in the later nymphal stages, and adults begin to appear toward the middle of April. By August adults become scarce, although they are found as late as November. In central Florida adults also occur in the winter months.

REFERENCES. Walker 1870–71, Rehn and Hebard 1916, Jago 1969.

Achurum minimipenne **Caudell**   Pl. 14

DISTRIBUTION. Southern Texas and eastern Mexico.

RECOGNITION. Male forewings usually slightly shorter than head plus pronotum; female forewings usually slightly longer than the pronotum. Males pale green or tan on the dorsum and brown or gray brown on the sides of the body. Females may be entirely pale brown, entirely pale green, green on the dorsum and brown on the sides, or brown on the dorsum and green on the sides. Hind knees with upper and lower pointed projections. Male subgenital plate shorter than hind femora. Body length to end of abdomen 19–30 mm in males, 25–34 mm in females.

A. minimipenne varies in robustness across its range. Plate 14 illustrates the extremes in shape, with the inland population being more robust than the coastal populations. The slenderest forms are found near Brownsville, Texas. In terms of increasing robustness, populations can be ranked as follows: Brownsville, Texas; Pueblo Viejo, Veracruz; Mesa de Llera, Tamaulipas; Ciudad Victoria, Tamaulipas.

HABITAT. Little about its biology is published, but judging from the body shape, this species probably clings to grasses.

LIFE CYCLE. In Texas adults have been collected from June to November and in Mexico from July to January.

REFERENCES. Caudell 1904, Jago 1969.

Achurum minimipenne

Genus PSEUDOPOMALA Morse

The position of this genus in the subfamily is problematical. Jago (1969) placed its only member, *P. brachyptera,* in the genus *Chloealtis,* but the differences in superficial morphology appear to me so great

that I have adopted the traditional classification. I place it with the genera that seem to be most similar, namely *Achurum* and *Mermiria*.
RECOGNITION. See *P. brachyptera*.

Pseudopomala brachyptera (Scudder)   Pl. 14

DISTRIBUTION. Mainly northern United States and western Canada but ranging southward through Kansas to Oklahoma.

RECOGNITION. Body color brown. Antennae strongly ensiform. Side of body without a white stripe as in *Paropomala* or *Prorocorypha*. Abdomen extending well beyond hind femora. Body sometimes with faint longitudinal stripes on head, pronotum, and forewings. Lateral pronotal carinae well developed, parallel, and cut by one sulcus. Lateral foveolae invisible from above. Fastigium of vertex divided into two concave areas by a median carina. Frontal costa grooved. Forewing length variable, usually not reaching ends of hind femora, but rarely extending to ends of hind femora. In females, forewings vary from being about as long as head plus pronotum to twice that length. Hind femora unicolorous. Hind tibiae with more than fifteen external spines. Male subgenital plate strongly pointed.

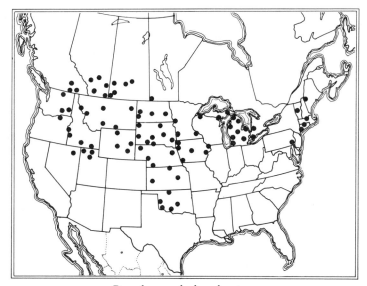

Pseudopomala brachyptera

HABITAT. Inhabits taller grasses through much of the prairie region and northeastern United States. Cantrall (1943) believed that a major original habitat was the tall grass prairies of the United States, and Morse (1896a: 382) wrote: "This peculiar locust is not uncommon locally on the coarser grasses found in waste lands, especially upon a species of bunch-grass (*Andropogon scoparius*)." Fox (1914) noted that in Pennsylvania and New Jersey it inhabits scrubby areas, usually in wooded surroundings, but is also associated with fringes of marsh elder, along the edges of salt marshes, and serpentine barrens.

BEHAVIOR. The song of *P. brachyptera,* easily heard in the field, is a series of ten to twenty individual leg strokes which produce sibilant *sh-sh-sh-sh* sounds of increasing intensity, delivered at about fifteen strokes per second (Otte 1970).

LIFE CYCLE. Adults from midsummer to fall.

REFERENCES. Scudder 1862, Morse 1896a, Blatchley 1920, Brooks 1958, Jago 1969, 1971.

Paropomala Genus Group

This group is composed of the genera *Paropomala, Prorocorypha,* and *Cordillacris.* The last genus is placed here only because of its superficial resemblance to *Paropomala virgata* and because it does not fit easily into any other group.

Genus PAROPOMALA Scudder

DISTRIBUTION. Dry grasslands in the southwestern United States and northern Mexico.

RECOGNITION. Slender, pale green or straw-colored insects with strongly ensiform antennae and with a ventral protuberance between the front pair of legs. Lateral foveolae on a vertical plane, not readily visible from above. Side of body with a whitish lower stripe or band and a gray brown postocular band. Hind femora without dark spots or bands. Pronotal disk without lateral carinae and with disk rounding gradually onto lateral lobes. Lateral margins of disk cut by two sulci. Posterior margin of disk rounded.

REFERENCES. Scudder 1899c, Rehn and Hebard 1906, 1908, Hebard 1927, 1929, 1935, Jago 1969.

• Identification of Paropomala Species

wyomingensis
1. Hind femora extending beyond ends of forewings
2. Male subgenital plate longer than front femora
3. Lower pale band on side of body much narrower than upper darker band
4. Abdomen extending beyond hind femora

pallida
1. Hind femora not extending beyond forewings
2. Male subgenital plate shorter than front femora
3. Lower pale band on side of body about as wide as upper dark band
4. Abdomen extending beyond hind femora

virgata
1. Hind femora not extending beyond forewings
2. Male subgenital plate shorter than front femora
3. Lower pale band on side of body about as wide as upper dark band
4. Abdomen not reaching end of hind femora

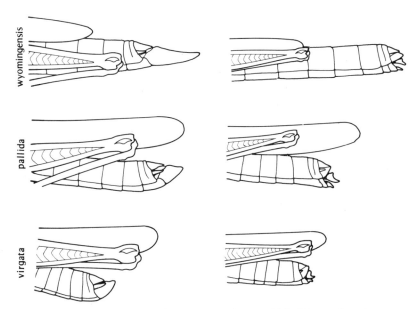

Fig. 74. Comparison of abdomen, wing, and hind femur lengths in *Paropomala* males (left) and females (right).

Paropomala wyomingensis (Thomas)   Pl. 14

DISTRIBUTION. Short-grass prairies from South Dakota to Mexico and south of a line running between southern California and northern Wyoming.

RECOGNITION. The following chracteristics distinguish this species from *P. pallida* and *P. virgata:* Hind femora extending beyond the forewings; abdomen extending beyond hind femora; pale band on side of body narrow (about as wide as thickness of front femora) and ending on base of middle leg; subgenital plate in males longer than front femora; arcuate groove on fastigium nearer to the back of fastigium than to the front. Body length to end of abdomen 19–31 mm in males, 25–40 mm in females.

HABITAT. Small to medium grass clumps in short-grass prairies and desert grasslands through much of the Great Plains and Rocky Mountain states. In deserts more common in taller grasses of depressions. It clings to grass stems with its short front legs and is difficult to flush. Upon escaping, it makes short flights and immediately seeks other clumps of grass.

LIFE CYCLE. Overwinters in the egg stage; adults from late June to October.

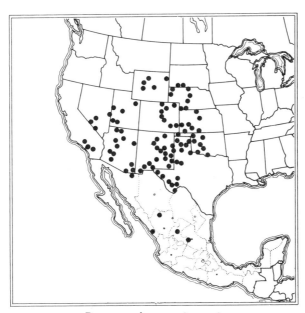

Paropomala wyomingensis

REFERENCES. Thomas 1871b, Rehn and Hebard 1906, Jago 1969, Otte 1970.

Paropomala pallida (Bruner)   Pl. 14

DISTRIBUTION. Found mainly west of a line between the Big Bend region of Texas and eastern Washington and ranging as far south as northern Durango, Mexico.

RECOGNITION. Distinguishable from *P. wyomingensis* and *P. virgata* by the following combination of characteristics: forewings and abdomen extending beyond hind femora; lower pale band on side of body about as wide as upper dark band and extending back onto hind femora; male subgenital plate slightly shorter than front femora; arcuate groove on the fastigium semicircular and at the anterior end passing through the middle or front half of the fastigium. Body length to end of forewings 16–25 mm in males, 23–40 mm in females.

HABITAT. Desert grasses throughout in southwestern United States. It is a good flyer and readily abandons grass clumps when disturbed. In southeastern Arizona it was found to feed on several

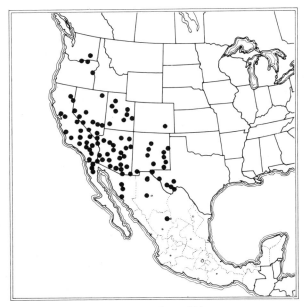

Paropomala pallida

grass species which grow as clumps and is common in clumps of *Muhlenbergia* under thorn shrubs and in *Aristida*. It does not live in tall and coarse grasses such as reeds and Johnson grass and also avoids low sparse grasses such as *Bouteloua aristidoides* and *B. gracilis*.

LIFE CYCLE. Overwinters in egg stage; nymphs common in early spring; adults from May to October depending on rainfall.

REFERENCES. Bruner 1904, Hebard 1935, Jago 1969, Otte and Joern 1977.

Paropomala virgata (Scudder)   Pl. 14

DISTRIBUTION. Colorado, New Mexico, and west Texas, and Chihuahua, Mexico.

RECOGNITION. Differs from other *Paropomala* species as follows: Forewings and hind femora both extend beyond end of abdomen; arcuate groove located near front of fastigium; often a small white horizontal streak directly above base of male forewings. Differs from *Cordillaris* in lacking spots on the side of the forewing and in possessing a protuberance between the front legs. Body length to end of forewings 16–21 mm in males, 23–30 mm in females.

HABITAT. Prairies and desert grasslands from Colorado to northern Mexico.

LIFE CYCLE. Adults from middle to late summer.

REFERENCES. Scudder 1899c, Hebard 1929, Jago 1969.

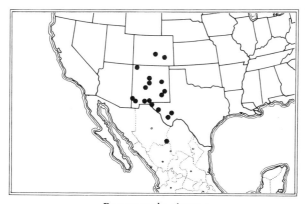

Paropomala virgata

Genus PROROCORYPHA Rehn

The genus is represented by a single species, which Jago (1969) assigned to the African genus *Mesopsis* Bolivar. The interpretation adopted here is that the genus represents an elongate member of a group that includes *Paropomala,* and the similarity to *Mesopsis* is the result of convergence.

RECOGNITION. See *P. snowi.*

REFERENCES. Jago 1969.

Prororcorypha snowi (Rehn)   Pl. 14

DISTRIBUTION. Known only from the Santa Rita, Tumacacori, Mule, and Huachuca Mountains of Arizona and the Sierra Saguarito, east of San Bernardo, Sonora, Mexico.

RECOGNITION. Exceedingly slender. Antennae ensiform and as long or longer than hind femur. Male subgenital plate more than twice as long as front femur. Fastigium of vertex convex and with a prominent median ridge. Forewings shorter than head plus pronotum. Body color mostly pale gray, pale green, yellowish, or pale brown, or a combination. Side of body with a pale white line running back from near base of antennae, beneath eye, along bottom of lateral lobes, and onto lower half of hind femora. Body length to end of abdomen 38–44 mm in males, 45–51 mm in females.

HABITAT. Tall grasses such as *Andropogon, Aristida,* and *Elyonurus* (Ball et al. 1942).

LIFE CYCLE. Adults from July to September.

REFERENCES. Rehn 1911, Ball et al. 1942, Jago 1969.

Prororcorypha snowi

Genus CORDILLACRIS Rehn

DISTRIBUTION. Western United States and prairie provinces of Canada west of a line running from western Minnesota to Big Bend, Texas.

RECOGNITION. Small gray and tan species, with a clearly defined gray brown postocular stripe to the back of the pronotum. Pronotal disk with triangular FDI at posterior lateral corners. Antennae pale, slightly ensiform. Forewings extending to end of abdomen but never beyond the hind femora. Lateral foveolae of vertex not visible from above. Fastigium of vertex concave, without a median carinula, and with the transverse groove arching near the front. Frontal costa of head slightly concave to strongly grooved. Pronotal carinae constricted in middle, well developed, and usually cut by one sulcus. Posterior margin of disk rounded. Compare with *Paropomala* (especially *P. virgata*), and *Opeia, Orphulella, Phlibostroma, Horesidotes,* and *Psoloessa.*

REFERENCES. Jago 1969.

• Identification of Cordillacris Species

occipitalis
1. Side of forewings with white streak
2. Front margin of lateral lobes without white spur intruding up into dark band

crenulata
1. Side of forewings without white streak
2. Front margin of lateral lobes with white spur intruding up into dark band

---

Cordillacris occipitalis (Thomas)   Pl. 13

DISTRIBUTION. Western half of United States and prairie provinces of Canada.

RECOGNITION. Separable from *C. crenulata* by the following characteristics: Side of forewing with a white streak between costa and subcosta directly above the base of the hind leg. Face without a dark streak descending from the eye to the front of the mandible. Front margin of lateral lobes without a white spur intruding up along the anterior margin into the upper dark band. Dark markings on forewings forming definite spots, not a wavy line. Hind tibiae usually with a tinge of orange. Body length to end of femora 14.8–22.0 mm in males, 17.0–26.0 mm in females.

HABITAT. The species lives in thinly grassed habitats where it runs about on the ground. In Arizona, Ball et al. (1942) reported that it occurs on shallow, sandy, or gravelly soil covered with scanty growth of short grasses and weeds, and ranges from the Upper Sonoran to the Transition zones from 4,000 to 9,000 feet.

BEHAVIOR. The male calling song is believed to attract females, but sexual pairing often occurs in the absence of stridulation. Males frequently raise and lower their hind femora as they walk about on bare patches of ground. At the same time the antennae are raised and lowered, touching the substrate. Courting males make similar movements as they approach females (Otte 1970).

LIFE CYCLE. Egg-overwintering. Adult season June to August in Arizona, July to September in Wyoming, mid-June to September in Kansas.

REFERENCES. Thomas 1873b, Ball et al. 1942, Brooks 1958, Jago 1969, Otte 1970.

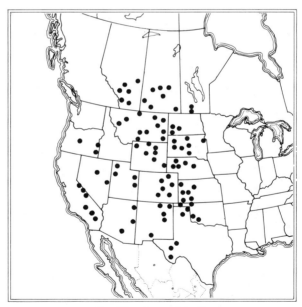

Cordillacris occipitalis

Cordillacris crenulata (**Bruner**)   Pl. 13

DISTRIBUTION. Western half of United States, except Pacific Northwest.

RECOGNITION. Separable from *C. crenulata* by the following features: Face with a dark streak between eye and front of mandible. Lateral lobes with a pale spur ascending into the dark band along the front margin. Lateral field of forewings without a pale streak directly above base of hind leg. Dark marks on forewings frequently forming a wavy line. Hind tibiae often becoming blackish at distal end. Body length to end of femora 13–17 mm in males, 16–23 mm in females.

HABITAT. The range overlaps broadly with that of *C. occipitalis*. The species live on the ground in thinly grassed areas. In Arizona it is one of the most common grasshoppers in drier areas where the soil is thin and the grass more parched, and it usually inhabits drier situations and lower elevations than *C. occipitalis*. In this state it is especially common on grama grass (*Bouteloua*) range.

BEHAVIOR. Mating behavior is similar to that described for *C. occipitalis* (Otte 1970).

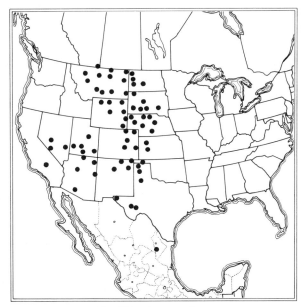

Cordillacris crenulata

LIFE CYCLE. Overwinters in the egg stage. Adult season, May to September in Arizona, late June to September in Colorado.

REFERENCES. Ball et al. 1942, Jago 1969, Otte 1970.

Acantherus Genus Group

The genus *Acantherus* is placed by itself because it does not fit easily into any other group.

Genus ACANTHERUS Scudder and Cockerell

This genus is represented by a single species. Its relationship to other North American gomphocerines is uncertain.

RECOGNITION. See *A. piperatus*.

REFERENCES. Scudder and Cockerell 1902, Ball et al. 1942.

Acantherus piperatus Scudder and Cockerell   Pl. 15

DISTRIBUTION. West Texas to southern Arizona and northwestern Mexico.

RECOGNITION. Body gray brown on top, yellowish on the sides. Top of lateral lobes with a dark band in the top third, yellowish in the lower two thirds. Antennae ensiform and more than one and a half times as the head plus pronotum. Lateral carinae constricted in the middle and cut by three sulci. Hind tibiae orange in distal two-thirds and banded with black and white in proximal third. Hind knees black. Hind femora faintly banded on the top face. Two basic color morphs are recognizable: (a) most individuals lack a median stripe and tend to be grayish on the head and pronotal disk; (b) rare individuals have a pale stripe running along the middle of the back from the front of the head to about the middle of the forewings. Color patterns of females are often quite variable within populations. Body length to end of femora 18–26 mm in males, 25–29 mm in females.

HABITAT. Arid, rocky hillsides and ridges from 2,000 to 4,500 feet in the Chihuahuan and Sonoran deserts. In Arizona, often found in medium to tall grasses under spiny shrubs and trees in saguaro-palo-verde deserts, but also ranges upward to the lower live-oak wood-lands.

LIFE CYCLE. In the southwestern United States adults occur from midsummer into fall. The species probably overwinters in the egg stage.

REFERENCES. Scudder and Cockerell 1902, Ball et al. 1942.

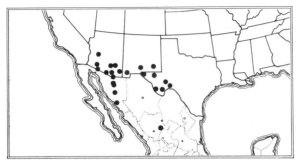

Acantherus piperatus

Acrolophitus Genus Group

This group includes the genera *Acrolophitus* and *Bootettix*. Both genera have pointed heads and saddle-shaped pronota, and both live and feed on dicotyledenous herbs and shrubs.

Genus ACROLOPHITUS Thomas

DISTRIBUTION. The four species of this genus range from north-ern Mexico through the U.S. mountain states to the prairies of west-ern Canada.

RECOGNITION. Head pointed, body hairy. Body color usually greenish, often mottled with dark green or dark gray. Side of head with green and pale stripes descending from the eye. Pronotum sad-dle-shaped, constricted on the prozona, and widening strongly on the metazona. Pronotum without prominent lateral carinae. Lateral carinal region cut by three sulci. Hind femora banded with green and ivory. Antennae long, filiform, and usually dark. Hindwings usually marked with dark pigmentation. All legs are relatively long.

REFERENCES. Jago 1969.

● **Identification of Acrolophitus Species**

hirtipes
1. Pronotum with high arching crest on metazona
2. Body mostly green; unspotted in *A. h. hirtipes,* spotted in *A. h. variegatus*
3. Hindwings green yellow with broad black or sooty band

maculipennis
1. Pronotum without crest
2. Body dark gray green or gray, mottled with black and milky white
3. Hindwings black

nevadensis
1. Pronotum without crest
2. Body green, unspotted
3. Hindwings yellowish in basal third, sooty gray in middle third

pulchellus
1. Pronotum without crest
2. Body green, spotted; similar to *A. h. variegatus*
3. Hindwings similar to *hirtipes*

Acrolophitus hirtipes (Say)   Pl. 15

DISTRIBUTION. *A. h. hirtipes* (Say) ranges from Oklahoma north to Alberta and Saskatchewan, and *A. h. variegatus* Bruner ranges from Kansas to San Luis Potosí, Mexico.

RECOGNITION. *A. h. hirtipes:* Pronotum with a high arching crest on the metazona. Hindwings with a broad dark band arching across the entire wing. Forewings and pronotum not striped or mottled, nearly unicolorous. Body length to end of wings 25–42 mm in males, 32–51 mm in females. *A. h. variegatus:* Very similar to *A. h. hirtipes* except that the forewings are strongly mottled, and the pronotum is distinctly banded.

HABITAT. *A. h. hirtipes* inhabits principally short-grass prairies and usually rolling country where the vegetation is sparse. According to Mulkern et al. (1969) it is the only prairie gomphocerine that is a forb-feeder. In Montana grasslands (Anderson and Wright 1952) and in the Canadian prairie provinces (Criddle 1933, Brooks 1958), it was found feeding on Boraginaceae. In Texas *A. h. variegatus* is often found on rocky soil in the Trans-Pecos shrub savanna. Isely (1937) found that it fed principally on the plant *Filago nivea,* and when this plant disappeared the population declined.

BEHAVIOR. *A. hirtipes* is unusual in possessing conspicuously banded hindwings, which suggests that some form of flight signal is used in mating behavior. Laboratory observations indicate that pair formation is also achieved by male stridulation. The song consists of a series of paired, short, ticking pulses (*t-t, t-t, t-t*). The number of pairs of pulses varies from twelve to eighteen, and courting males produce the same sound (Otte 1970).

LIFE CYCLE. In Texas juveniles are encountered in March and April; adults are common in May and June, rare in July and August. The early appearance of juveniles and adults suggests that eggs hatch in the fall and nymphs are the overwintering stage (Isely 1937). In Colorado, Montana, and Alberta adults appear to be most abundant in July and August.

REFERENCES. Bruner 1904, Tinkham 1948, Brooks 1958, Jago 1969, Otte 1970.

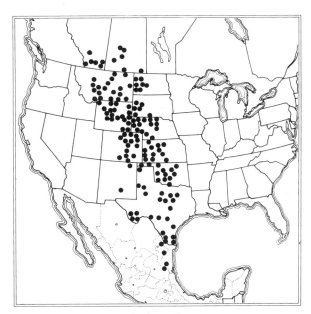

Acrolophitus hirtipes

Acrolophitus maculipennis (**Scudder**)   Pl. 15

DISTRIBUTION. Durango and San Luis Potosí, Mexico, north to west Texas, southern New Mexico, and southern Arizona. Never abundant.

RECOGNITION. A strongly mottled gray-green and ivory insect. Forewings mottled gray and black. Hindwings black. Antennae long and mostly black. Face and pronotum banded with gray-green and ivory. Hind femora either green and ivory or black and ivory. Metazona without a prominent crest. Body length to end of wings 23–35 mm in males, 28–43 mm in females.

HABITAT. Creosote desert and desert grassland. Feeds principally on a low mat-like plant, *Coldenia canescens* (Tinkham 1948). Inhabits arid regions from Arizona and New Mexico to central Mexico. In Arizona it also occurs in low rocky hills of the Lower Sonoran Zone (Ball et al. 1942).

BEHAVIOR. Mating behavior is largely unknown, but the black color of the wings suggests that the males use the wings in a flight or courtship display. The song consists of a repeated series of three ticks. From fourteen to twenty-four such bursts have been recorded (Otte 1970).

LIFE CYCLE. In the United States adults appear from June through September.

REFERENCES. Scudder 1890, Ball et al. 1942, Tinkham 1948, Jago 1969, Otte 1970.

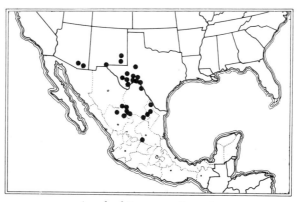

Acrolophitus maculipennis

Acrolophitus nevadensis (**Thomas**)   Pl. 15

DISTRIBUTION. A rare species ranging from east central Arizona to southwestern Utah and southern Nevada. In Arizona it has been found 8 and 10 miles NE of Flagstaff, on the Schultz Pass road in

Coconino County and near Springerville. In Utah it has been collected in Wayne County, Asay Bench, and Paunsagunt Plateau SE of Bryce (or Red) Canyon in Garfield County.

RECOGNITION. Body almost entirely green. Forewings unspotted. Hindwings with a dark area in the central part, but not as prominent as in *A. hirtipes*. Metazona without a prominent crest. Pronotum usually with a pink or ivory line running along the lateral edges of the disk. Front and middle legs often with pink coloration. Body length to end of forewings 22–29 mm in males, 29–36 mm in females.

HABITAT. White and Nickerson (1951) found *A. nevadensis* on *Lappula coronata* in clearings in a *Pinus ponderosa* forest. Ball et al. (1942) reported it associated with *Actinia richardsoni* and feeding upon *Artemisia frigida*, *Gutierrezia* sp., and perhaps also *Bouteloua* sp. Alexander and Rodeck (1952) collected three individuals at the edge of the sagebrush area on Round Top in Dinosaur National Monument in Colorado. White and Nickerson (1951) have reported on an analysis of *A. nevadensis* chromosomes.

LIFE CYCLE. All adults in the Philadelphia Academy collection were collected from late July through late August.

REFERENCES. Thomas 1873c, Ball et al. 1942, Jago 1969.

Acrolophitus pulchellus (**Bruner**)   Pl. 15

DISTRIBUTION. Only two specimens of *A. pulchellus* exist in museums, both collected at Birch Creek, Idaho.

RECOGNITION. Jago (1969) considered *A. pulchellus* to be a syn-

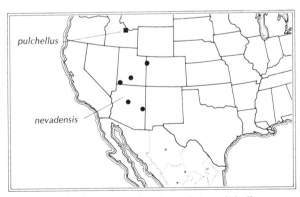

Acrolophitus nevadensis and *A. pulchellus*

onym of *A. nevadensis*, but there are two reasons for thinking that it is specifically distinct: (1) The distributions of *A. pulchella* and *A. nevadensis* are separated by more than 400 miles; and (2) in *A. pulchella* the forewings are conspicuously spotted, the hindwings are more strongly banded, and the patterns of body markings on the head and pronotum are different. Body length to end of forewings: 30 mm in one female.

HABITAT. The only existing specimens were found associated with the plant *Grayia polygaloides*.

LIFE CYCLE. The specimens were collected in August 1883.

REFERENCES. Rehn and Hebard 1912a, Jago 1969.

Genus BOOTETTIX Bruner

DISTRIBUTION. Sonoran and Chihuahuan deserts. The members of the genus inhabit only creosote bushes (*Larrea divaricata*).

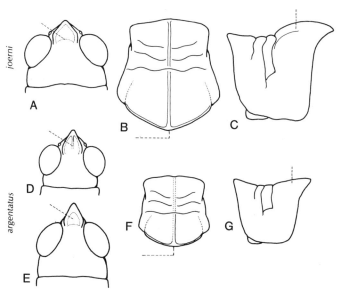

Fig. 75. *Bootettix*. A–C, *B. joerni* male from near Cuatrociénegas, Coahuila, Mexico; D, F, G, *B. argentatus* male from near Cuatrociénegas; E, *B. argentatus* from southern California.

RECOGNITION. Body color olive green. Pronotal disk strongly constricted on prozona (Fig. 75). Pronotum and side of thorax with shiny, pearly white markings. Hind femora also banded with white. Head pointed.

REFERENCES. Jago 1969.

Bootettix argentatus **Bruner**   Pl. 15

DISTRIBUTION. Widespread through the deserts of southwestern United States and northern Mexico.

RECOGNITION. See *B. joerni* and Fig. 75.

HABITAT. The color of this grasshopper, which lives exclusively on creosote bushes, blends well with the shiny foliage, making them extremely difficult to locate. When a bush is heavily disturbed, individuals will jump or fly out of it and immediately seek the same bush by flying in a short circle or fly to nearby bushes. Nymphs that land on the ground quickly jump toward a bush. Population densities may reach high levels, with as many as thirty individuals residing on a small bush. At high densities individuals are likely to fly when a bush is approached, but at low densities they may sit still until touched.

Males call day or night, and stridulation consists of two sounds, a short *zick* or *tick* followed by a longer *zzzzzzt*. Similar sounds are produced by courting males (Otte, 1970).

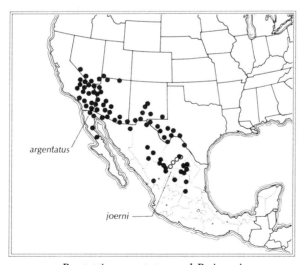

Bootettix argentatus and *B. joerni*

LIFE CYCLE. Adults occur at all seasons but are most abundant from September to December.

REFERENCES. Jago 1969, Otte 1979b.

Bootettix joerni Otte   Pl. 15

DISTRIBUTION. Known only from the region of Torreon and Cuatrociénegas, north central Mexico, on creosote bushes.

RECOGNITION. In the region of geographic overlap, *B. joerni* differs from *B. argentatus* as follows: Hindwings pink (clear or bluish in *B. argentatus*); fastigium relatively short (Fig. 75) (relatively long in *B. argentatus*); fastigium usually without a small median carinula (*B. argentatus* with a carinula); metazona humped (flattish in *B. argentatus*) transverse arcuate groove of fastigium usually distinct (usually indistinct or absent in *B. argentatus*) lateral foveolae usually visible from above (usually invisible in *B. argentatus*); hind margin of pronotal disk slightly angulate (rounded in *B. argentatus*)

LIFE CYCLE. Adults have been collected in October and November.

REFERENCES. Otte 1979b.

Stethophyma Genus Group

This aberrant group includes the genera *Stethophyma,* known from Asia and America, and *Mecostethus* from Eurasia. I tentatively place it with the Gomphocerinae, principally because it resembles this group in behavior and appearance. On the basis of the stridulatory apparatus, it should be grouped with the Oedipodinae.

Genus STETHOPHYMA Fisher

TAXONOMY. In the past there has been some difficulty concerning the names *Mecostethus* Fieber (1852) and *Stethophyma* Fisher (1853); see Jago (1971) for the history of this matter. The scheme adopted in this work is based on taking *Gryllus* (*Locusta*) *grossus* L. as the type species of *Stethophyma,* a species clearly congeneric with all of the North American species discussed here, and taking *Gryllus alliaceus* Germar as the type species of *Mecostethus.*

DISTRIBUTION. The genus *Stethophyma* is Palaearctic and Nearctic in distribution. Eurasia has three species: *S. grossum* (L.), *S. tsherakii* (Ikonn.), and *S. magister* (Rehn), and North America has three species: *S. lineata* (Scudder), *S. celata* Otte, and *S. gracile* (Scudder). *S. lineata,* which ranges across the entire North American continent from Alaska to Newfoundland, is very similar to *S. grossum,* and the two may even belong to a single species.

RECOGNITION. Body color yellow, yellow brown, green, or a combination. Greatest body length: males more than 25 mm, females more than 30 mm. Hind femora in males without stridulatory pegs; instead, males possess a raised, intercalary vein between the M and Cu veins on the forewings, which is used as the file in stridulation. Females with long ovipositor valves. Antennae filiform. Forewings always extending beyond hind femora in males; in females forewings sometimes extending beyond end of abdomen, sometimes not quite reaching the end. Frontal ridge usually grooved below median ocellus and convex near top. Lateral foveolae are small and triangular. Fastigium concave or flat, with straight anterior ridges, and with a median carina. Disk of pronotum without FDI, with nearly parallel sides on the prozona, and becoming much wider on metazona (except in *S. celata* where the lateral carinae are nearly parallel). Lateral carinae distinct and cut by one, two, or three sulci. Hind knees black. Top and outer faces of hind femora not banded. Hind tibiae yellow, sometimes banded with black. Venter of abdomen yellow.

REFERENCES. Scudder 1862, Bei-Bienko and Mishchenko 1964, Jago 1971.

• **Identification of Stethophyma Species**

lineata
1. Spines on hind tibiae entirely black
2. Posterior width of pronotum more than one and a half times anterior width
3. Side of forewings with white streak
4. Lateral carinae cut by three sulci
5. Lateral lobes without dark band in upper third

gracile
1. Spines on hind tibiae entirely black
2. Posterior width of pronotum more than one and a half times anterior width
3. Side of forewings without white streak

4. Lateral carinae cut by one or two sulci
5. Lateral lobes without dark band in upper third
celata
1. Spines on hind tibiae yellow, tipped with black
2. Posterior width of pronotal disk less then one and a half times anterior width
3. Side of forewings without white streak
4. Lateral carinae cut by one sulcus
5. Lateral lobes with dark band in upper third

***Stethophyma lineata* (Scudder)**   Pl. 16

DISTRIBUTION. Ranges in wide arc from Alaska to Newfoundland, generally parallel to the major North American lakes.

RECOGNITION. Differs from *S. celata* and *S. gracile* as follows: Side of forewings always with a white or pale horizontal streak directly above base of hind leg. Lateral carinae cut by three sulci.

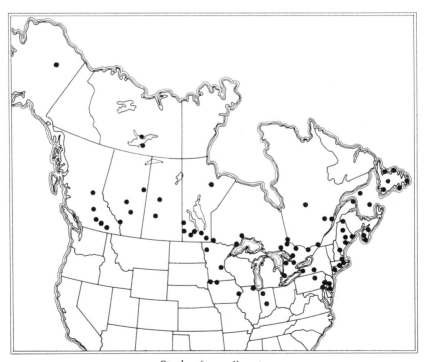

Stethophyma lineata

Body length 26–34 mm to end of wings in males, 34–41 mm to end of abdomen in females.

HABITAT. Low boggy meadows, swamps, marshes, and edges of lakes and tamarack bogs.

LIFE CYCLE. Adults from July to September in the northern United States.

REFERENCES. Scudder 1862, Blatchley 1920, Brooks 1958.

Stethophyma gracile (Scudder)   Pl. 16

DISTRIBUTION. Ranges across southern Canada and northern United States, extending south in the mountain states to Colorado.

RECOGNITION. Differs from *S. lineata* in lacking a white streak on the side of forewings and in having lateral carinae cut by one or two sulci. It differs from *S. celata* in that lateral carinae diverge strongly on prozona so that the posterior width is more than one and a half times the anterior width and in having completely black spines on the hind tibiae. Body length 25–29 mm to end of wings in males, 30–37 mm to end of abdomen in females.

HABITAT. According to Morse (1896a), it inhabits wet sedgy

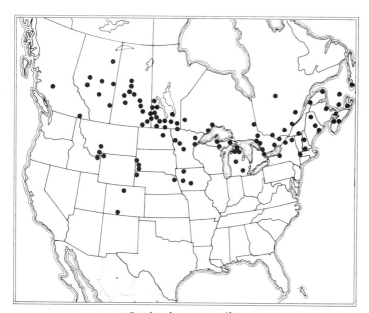

Stethophyma gracile

meadows, bushy swamps, and even mountain tops. "On Greylock it is common in the low grasses and bushes of the extreme summit and on Mt. Washington in a sedgy area." In Michigan Hebard found it in very thick high grass in a clearing.

BEHAVIOR. Males stridulate very loudly and can be heard from many meters away.

LIFE CYCLE. Adults from July to September.

REFERENCES. Scudder 1862, Blatchley 1920, Brooks 1958.

Stethophyma celata **Otte**   Pl. 16

DISTRIBUTION. This species appears to have two main centers of distribution, one in the prairie states and another along the eastern United States. In the east it is now known only from Connecticut, Massachusetts, and South Carolina. Eastern populations differ from western ones in several respects (see below).

RECOGNITION. Differs from *S. lineata* in lacking a white stripe on side of forewings, and in having top third of lateral lobes dark brown to black. It differs from *S. gracile* in having yellow posterior tibial spines with black tips and in having the upper side of the lateral lobes dark. In western states it differs from both *S. gracile* and *S. lineata* in having yellow instead of red lower marginal area on the hind femora. Body length 29–33 mm to end of wings in males, 39–47 mm to end of abdomen in females.

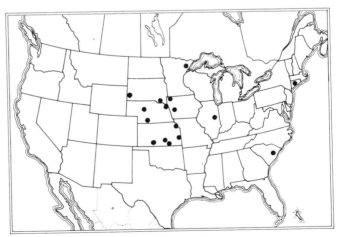

Stethophyma celata

HABITAT. Tall grasses, moist swales, and marshes. In Minnesota Somes (1914) found it in dense tangles of tamarack swamps. In Connecticut it was found in marshes with *S. lineata* (Morse 1896a).

LIFE CYCLE. Adults in July and August.

REFERENCES. Scudder 1862, Blatchley 1920, Otte 1979b.

Acridinae

In the New World this subfamily is best represented in South America; only *Metaleptea* and *Orphula* extend north of the Panama Canal. The subfamily is evidently represented by a single tribe, the Hyalopterygini, which Dirsh (1975) raised to subfamily status (Hyalopteryxinae).

RECOGNITION. The New World Acridinae (all Hyalopterigini) share with the Gomphocerinae most of the characteristics listed for that group but can readily be distinguished on the basis of three characteristics: (1) The hind tibiae in males always lack stridulatory pegs, but this feature alone is not sufficient to separate the two subfamilies because some gomphocerine species (some *Orphulella* species, *Melanotettix,* and southern populations of *Achurum carinatum*) have secondarily lost their pegs; (2) the forewings of both sexes are obliquely truncated at the apex (Fig. 21B), although in females of *Orphula* this feature is sometimes barely discernible; and (3) the hindwings of males always have enlarged cells. In some species (*Orphula*) the enlargement is moderate, while in others (*Hyalopteryx*) the cells are huge and rectangular. Some species of Gomphocerinae also have enlarged hindwing cells (*Orphulella tolteca, O. orizabae,* and *Phaneroturis* species), but these species all possess stridulatory pegs.

The enlarged cells and the thickened leading cells on the forewings are associated with sound produced during flight displays. Such crepitation is similar to that of the Oedipodinae, but some species (*Hyalopteryx*) produce a more musical tone.

Hyalopteryx Genus Group

This group includes the following genera: *Hyalopteryx* Charpentier, *Paulacris* Rehn, *Guaranacris* Rehn, *Cocytotettix* Rehn, *Allo-*

truxalis Rehn, *Eutryxalis* Bruner, *Parorphula* Bruner, *Metaleptea* Brunner, and *Orphula* Stål. In this work the names *Cumarala* Hebard, *Sisantum* Bruner, and *Thyriptilon* Bruner are treated as junior synonyms of *Orphula* Stål. Rehn (1944: 182), who called the group the Hyalopteryges, notes: "It seems probable that the generic differentiation of the group took place in the southern tropical and subtropical portions of South America, where all the genera now occur, and that the penetration of North America has been relatively recent, dating very probably from the re-establishment of the Panama land bridge in the Pliocene." This is the only group of New World Acridinae. See Subfamily Recognition above.

• **Identification of North American Genera**

Metaleptea
　　1. Hind knees with pointed upper and lower lobes
　　2. Front margin of fastigium rounded
　　3. Extreme end of forewings pointed
　　4. Subgenital plate pointed at end
Orphula
　　1. Hind knees without pointed lobes
　　2. Front margins of fastigium with straight sides, angulate
　　3. Extreme end of forewings rounded
　　4. Subgenital plate not pointed at end

Genus METALEPTEA Brunner de Wattenwyl

DISTRIBUTION. Eastern United States to Argentina. This genus and *Orphula* are the only members of the Hyalopterigini which have penetrated North America.

RECOGNITION. Lateral foveolae usually absent or, when present, hidden beneath the ridges of the fastigium. Frontal ridge with a deep V-shaped groove. Fastigium of vertex with rounded sides and a prominent arcuate groove. Antennae ensiform. End of forewings obliquely truncated and with distal extremity pointed. Disk of pronotum usually with nearly parallel sides. Hind knees with pointed upper and lower lobes; lower lobes usually more pointed. Subgenital plate pointed at end.

REFERENCES. Giglio-Tos 1897, Gurney 1940.

Metaleptea brevicornis (**Johannson**)   Pl. 16

DISTRIBUTION. Eastern United States south to Argentina.

RECOGNITION. Differs from *Orphula* mainly in the shape of the subgenital plate and the shape of the fastigium. Forewings always extend well beyond hind femora. Lateral foveolae absent and foveolar area hidden beneath the fastigial ridges. Fastigium with a median ridge that does not extend back of the fastigial area. Lateral pronotal carinae distinct and nearly parallel and cut by two sulci. Hind femora not banded. Upper and lower lobes of hind knees pointed posteriorly. Subgenital plate of males pointed (more rounded in *Orphula*). Males usually pale to dark brown on the side of the body and green on dorsum. In females, top and sides may be similarly colored, varying from pale brown to green, or they may be light brown on top and green on the sides. Body length to end of wings 25–38 mm in males, 36–53 mm in females.

HABITAT. Marshes, tall wet grasses in upland depressions. Blatchley (1920) described the habitat as follows: "It frequents only the tall grasses and sedges along the margins of lakes, ponds, streams and swales, and in such localities is usually locally abun-

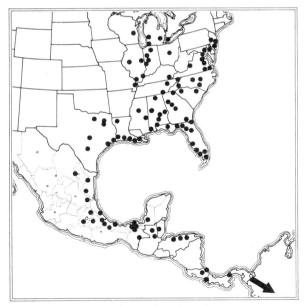

Metaleptea brevicornis

dant." In New Jersey the species is found in both fresh- and salt-marsh areas but is more abundant in the latter.

LIFE CYCLE. Adults have been recorded from about the middle of July to September in Indiana. In the southeastern states adults are found from July to October. In tropical regions adults are found at all seasons.

REFERENCES. Rehn and Hebard 1916, Blatchley 1920.

Genus ORPHULA Stål

DISTRIBUTION. Eastern Mexico to Argentina.

RECOGNITION. Differs from *Orphulella* in lacking thickened front and middle femora in males and in having the posterior margin of the forewing angulate (Fig. 76). Differs from *Metaleptea* as follows: Hind knees without pointed upper and lower lobes; antennae only slightly ensiform, sometimes simply flattened in the basal segments; lateral foveolae present but small; ridges of the fastigium not over-hanging the foveolae; front margins of fastigium with straight sides; frontal costa flat or shallowly grooved; disk of pronotum becoming much wider on the metazona; end of forewings obliquely cut, but posterior extremity rounded (pointed in *Metaleptea*). Subgenital plate not strongly pointed.

REFERENCES. Stål 1873, Rehn 1916, 1917.

• Identification of Orphula Species

azteca
 Area of forewings between cubitus and postcubitus with large cells

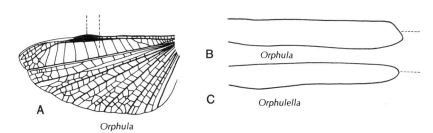

Fig. 76. *Orphula*. A, hindwing of *O. azteca*. B, C, comparison of forewings of *Orphula* and *Orphulella*.

vitripenne
  Area of forewings between cubitus and postcubitus with very
  small cells

Orphula azteca (**Saussure**)   Pl. 16

DISTRIBUTION. Tamaulipas and San Luis Potosí, Mexico, southward to northern South America.

RECOGNITION. Differing from *O. vitripenne* as follows: Triangular area of forewing between the cubitus and postcubitus veins with
large cells (small cells in *O. vitripenne*). Dorsum green or pale
brown, side of body dark brown, dark gray or blackish. Pronotal
disk not nearly as narrow as in *O. vitripenne*. *O. azteca* is superficially similar to *Orphulella walkeri* and *Chiapacris velox* and should
be compared with them. Body length to end of wings 16–24 mm in
males, 21–32 mm in females.

HABITAT. Grassy clearings, forest openings, and forest margins.

LIFE CYCLE. Adults are found throughout the year in Central
America and have been collected between July and November in
eastern Mexico.

REFERENCES. Saussure 1861, Bruner 1904.

Orphula vitripenne (**Bruner**)   Pl. 16

DISTRIBUTION. Yucatán Peninsula.

RECOGNITION. Differs from *O. azteca* as follows: Body color
chocolate brown, but usually lighter on dorsum; pronotal disk relatively narrow on prozona; triangular area of forewing between cu-

Orphula azteca and *O. vitripenne*

bitus and postcubitus with many small cells. Body length to end of wings 24–27 mm in males, 31–34 mm in females.

HABITAT. Rice collected this species in a "tall dry forest."

LIFE CYCLE. Adults in the Philadelphia Academy collection were all taken in July. For the remaining specimens no date is indicated.

REFERENCES. Bruner 1904.

Appendixes

Glossary

References

Taxonomic Index

Taxonomic Changes
Made in This Volume

This list includes all new synonyms and generic reassignments adopted in this work (see Appendixes 2 and 3 for species and generic nomenclature).

Achurum hilliardi Gurney, new synonym of **A. sumichrasti** (Saussure)

Acrolophitus maculipennis (Scudder), new combination for *Pedioscertetes maculipennis*

Acrolophitus nevadensis (Thomas), new combination for *Pedioscertetes nevadensis*

Acrolophitus pulchella (Bruner), new combination for *Pedioscertetes pulchella*

Ageneotettix brevipennis (Bruner), new combination for *Zapata brevipennis*

Ageneotettix curtipennis Bruner, new synonym of **Ageneotettix deorum** (Scudder)

Ageneotettix salutator (Rehn), new combination for *Zapata salutator*

Amblytropidia bruneri Hebard, new synonym of **Amblytropidia elongata** Bruner

Amblytropidia costaricensis Bruner, new synonym of **Amblytropidia mysteca** (Saussure)

Amblytropidia insignis Hebard, new synonym of **Amblytropidia trinitatis** Bruner

Amblytropidia magna Bruner, new synonym of **Amblytropidia trinitatis** Bruner

Amblytropidia occidentalis Saussure, new synonym of **Amblytropidia mysteca** (Saussure)

Amblytropidia pulchella Hebard, new synonym of **Amblytropidia trinitatis** Bruner

Amphitornus carinatus Scudder, new synonym of **Chloealtis abdominalis** (Thomas)

Astehelius Vickery, new synonym of **Cibolacris** Hebard

Aulocara parallelum Scudder, new synonym of **Aulocara elliotti** (Thomas)

Boopedon hoagi Rehn, new synonym of **Boopedon nubilum** (Say)

Cibolacris crypticus (Vickery), new combination for *Astehelius crypticus*

Cumarala Hebard, new synonym of **Orphula**

Eritettix obscurus (Scudder), new combination for *Amphitornus obscurus* (formerly *Macneillia obscura*)

Eritettix variabilis Bruner, new synonym of **Eritettix simplex** (Scudder)

Macneillia Scudder, new synonym of **Eritettix** Bruner

Opeia mexicana Bruner, new synonym of **Opeia obscura** (Thomas)

Orphula vitripenne (Bruner), new combination for *Thyriptilon vitripenne*

Oxycorphus mexicanus Saussure, new synonym of **Orphula azteca** Saussure

Pedioscirtetes Thomas, new synonym of **Acrolophitus** Thomas

Psoloessa brachyptera (Bruner) new combination for *Stirapleura brachyptera*

Psoloessa salina (Bruner) new combination for *Stirapleura salina*

Psoloessa thamnogaea Rehn, new synonym of **Psoloessa texana** Scudder

Sisantum Bruner, new synonym of **Orphula** Stål

Stenobothrus alticola Rehn, new synonym of **Stenobothrus shastanus** (Scudder)

Stenobothrus sordidus, new synonym of **Stenobothrus shastanus** (Scudder)

Stirapleura Scudder, new synonym of **Psoloessa** Scudder

Syrbula fuscovittata Thomas, new synonym of **S. montezuma** (Saussure)

Syrbula pacifica Bruner, new synonym of **S. montezuma** (Saussure)

Thyriptilon Bruner, new synonym of **Orphula** Stål

Zapata Bruner, new synonym of **Ageneotettix** McNeill

Zapata bucculenta Rehn, new synonym of **Ageneotettix brevipennis** (Bruner)

The Genera of
Gomphocerinae and Acridinae

Each genus of Gomphocerinae and Acridinae is listed in bold type, with its synonyms in italics, along with the reference to the original description of the genus and the name of the type species.

Acantherus Scudder and Cockerell 1902: 22. Type species: *Acantherus piperatus* Scudder and Cockerell, by original designation.

Achurum Saussure 1861: 313. Type species: *Truxalis sumichrasti* Saussure. Subsequent designation by McNeill 1897b: 202

　　Radinotatum McNeill 1897b: 199–201. Type species: *Truxalis brevipenne* Thomas, by original designation. Synonymized by Jago 1969.

Acrolophitus Thomas 1871a: 278. Type species: *Gryllus hirtipes* Say, by original designation.

　　Pedioscirtetes Thomas 1873c: 295. Type species: *Pedioscirtetes nevadensis* Thomas, by monotypy. NEW SYNONYM.

　　Acrocara Scudder 1890: 436. Type species: *Acrocara maculipenne* Scudder, by monotypy. Synonymized by Hebard 1926: 50.

Aeropedellus Hebard 1935: 187. Type species: *Gomphocerus clavatus* Thomas, by original designation.

Ageneotettix McNeill 1897a: 71. Type species: *Chrysochraon deorum* Scudder, subsequent designation by Kirby 1910.

　　Eremnus McNeill 1897b: 268. Homonym of *Eremnus* Schonh.

　　Zapata Bruner 1904: 102. Type species: *Zapata brevipennis* Bruner, by original designation. NEW SYNONYM.

Amblytropidia Stål 1873: 93. Type species: *Amblytropidia ferruginosa* Stål, subsequent designation by Rehn 1904.

Amphitornus McNeill 1897b: 223. Type species: *Stenobothrus coloradus* Thomas, subsequent designation by Kirby 1910.

　　Akentetus McNeill 1897b: 225. Type species: *Akentetus unicolor* McNeill, by monotypy.

Aulocara Scudder 1876a: 266. Type species: *Aulocara caerulipes* Scudder, subsequent designation by Scudder 1897; synonym of *A. elliotti* Thomas.

Oedocara Scudder 1876b: 289. Type species: *O. strangulatum* Scudder, by monotypy. Synonymized by Bruner 1905: 110.

Coloradella Brunner de Wattenwyl 1893: 123. Type species unknown. Brunner introduces the name in key and states in a footnote that two species, which he does not name, belong to the genus. Bruner 1897: 129 uses the same *Coloradella? brunnea* Thomas to refer to *Stenobothrus brunneus* Thomas. Kirby 1910 also includes only this species in his catalogue. Synonymized by Bruner 1905: 110.

Drepanopterna Rehn 1927: 226. Type species: *Aulocara femoratum* Scudder, by original designation. Synonymized by Jago 1971.

Boopedon Thomas 1870: 83. Type species: *Gryllus nubilus* Say, subsequent designation by Jago 1971: 263.

Morseiella Hebard 1925a: 273. Type species: *Boopedon flaviventris* Bruner, by original designation. Synonymized by Jago 1971.

Bootettix Bruner 1890: 58 (February). Type species: *Bootettix argentatus* Bruner, by original designation.

Gymnes Scudder 1890: 434 (December). Type species: *Gymnes punctatus* Scudder, by original designation. Synonymized by Bruner 1904: 52.

Chiapacris Otte 1979b. Type species: *Chiapacris velox* Otte, by original designation.

Chloealtis Harris 1841: 160. Type species: *Locusta (Chloealtis) conspersa* Harris, subsequent designation by Kirby 1910: 126.

Neopodismopsis Bei-Bienko 1932: 56. Type species: *Chrysochraon abdominalis* Thomas, by monotypy. Synonymized by Jago 1969.

Napaia McNeill 1897b: 212. Type species: *Napaia gracilis,* by monotypy. Synonymized by Jago 1969.

Oeonomus Scudder 1899b: 47. Type species: *Oeonomus altus* Scudder, by monotypy.

Chorthippus Fieber 1852: Type species: *Acrydium albomarginatum* DeGeer, by subsequent designation. See Vickery 1964, 1967.

Chrysochraon Fisher de Waldheim 1853. Type species: *Podisma dispar* Germar, subsequent designation by Kirby 1910.

Barracris Gurney, Strohecker, and Helfer 1964. Type species: *Barracris petraea* G, S and H, by original designation. Synonymized by Jago 1971. See Jago 1971: 250 for complete generic synonymy.

Cibolacris Hebard 1937: 268. Type species: *Oedipoda parviceps* Walker, by original designation.

Astehelius Vickery 1969a: 1223. Type species: *Astehelius crypticus* Vickery, by original designation. NEW SYNONYM.

Compsacrella Rehn and Hebard 1938: 209. Type species: *C. poecila* Rehn and Hebard, by monotypy.

Cordillacris Rehn 1901. Type species: *Stenobothrus occipitalis* Thomas, by original designation.

Alpha Brunner de Wattenwyl 1893: 121. Mentioned in key and footnote, preoccupied by *Alpha* Saussure.

Dichromorpha Morse 1896: 326. Type species: *Chloealtis viridis* Scudder, by original designation.

Clinocephalus Morse 1896: 326. Type species: *Clinocephalus elegans* Morse, by original designation. Synonymized by Otte 1979a.

Eritettix Bruner 1890: 56. Type species: *Eritettix variabilis* Bruner, subsequent designation by Rehn 1907: 33, new synonym of *E. simplex*.

Pedeticum McNeill 1897b: 216. Homonym of *Pedeticus* Laporte, Scudder 1898: 325.

Mesochloa Scudder 1898: 231. Type species: *Eritettix abortivus* Bruner, by original designation. Synonymized by Jago 1971: 246.

Macneillia Scudder 1898: 231. Type species: *Chrysochraon obscurus* Scudder, by monotypy. NEW SYNONYM.

Eupedetes Scudder and Cockerell 1902: 24. Type species: *E. carinatus* Scudder 1902, by monotypy. Synonymized by Rehn 1907 under *Eritettix variabilis* Bruner.

Esselenia Hebard 1920: 71. Type species: *Esselenia vanduzeei* Hebard, by original designation.

Eupnigodes McNeill 1897b. New name for the homonym *Pnigodes* Leconte. Type species: *Pnigodes megacephala* McNeill, by monotypy.

Heliaula Caudell 1915: 27. Type species: *Aulocara rufa* Scudder, by original designation.

Horesidotes Scudder 1899b: 49. Type species: *Horesidotes cinereus* Scudder, by monotypy.

Leurohippus Uvarov 1940. New name for *Leurocerus* Bruner 1911. Type species: *Coccytotettix linearis* Rehn 1906a: 374. Designation by Bruner 1911: 23.

Caribacris Rehn and Hebard 1938: 201. Type species: *Amblytropidia stoneri* Caudell 1922: 32, by original designation. Synonymized by Jago 1971.

Ligurotettix McNeill 1897b: 257. Type species: *Ligurotettix coquilletti* McNeill, by monotypy.

Goniatron Bruner 1905: 112. Type species: *Goniatron planum* Bruner, by monotypy. Synonymized by Jago 1971.

Melanotettix Bruner 1904: 90. Type species: *Melanotettix dibelonius* Bruner, by monotypy.

Mermiria Stål 1873: 102. Type species: *Mermiria belfragii* Stål, by original designation, synonym of *Mermiria picta* (Walker).

Papagoa Bruner 1902: 27. Type species: *Papagoa arizonensis* Bruner, by original designation.

Metaleptea Brunner de Wattenwyl 1893: 118. Type species: *Gryllus brevicornis* Johannson. Designation by Bruner 1895: 58–69. See Gurney 1940: 92 for details of designation.

Opeia McNeill 1897b. Type Species: *Oxycoryphus obscurus* Thomas, by monotypy.

Orphula Stål 1873. Type species: *Truxalis pagana* Stål, by monotypy.

Sisantum Bruner 1904: 68. Type species: *Sisantum notochloris* Bruner, by monotypy. NEW SYNONYM.

Thyriptilon Bruner 1904: 69. Type species: *Thyriptilon vitripenne* Bruner, by monotypy. NEW SYNONYM.

Cumarala Hebard 1923: 202. Type species: *Cumarala annectens* Hebard, by monotypy. NEW SYNONYM.

Orphulella Giglio-Tos 1894: 10. Type species: *Stenobothrus gracilis* Scudder, subsequent designation by Rehn 1904 (synonym of *Orphulella speciosa*). Synonymized under *Orphulina* by Jago 1971.

Linoceratium Bruner 1904: 84. Type species: *Linoceratium boucardi* Bruner, by monotypy. Synonymized by Rehn 1916.

Parachloebata Bruner 1904: 83. Type species: *Parachloebata pratensis* Bruner, by monotypy. Synonym of *P. scudderi* (Bolivar). Synonymized by Otte 1979a.

Isonyx Rehn 1906b: 36. Type species: *Isonyx paraguayensis* Rehn, by monotypy. Synonymized by Otte 1979a.

Orphulina Giglio-Tos 1894: 9. Type species: *Orphulina pulchella* Giglio-Tos, by original designation.

Paropomala Scudder 1899c. Type species: *Mesops cylindricus* Bruner, subsequent designation by Rehn and Hebard 1912a. Synonym of *P. wyomingensis* (Thomas).

Eremiacris Hebard 1929: 437. Type species: *Paropomala virgata* Scudder, by original designation. Synonymized by Jago 1969.

Phaneroturis Bruner 1904: 88. Type species: *Phaneroturis cupido* Bruner, by monotypy.

Phlibostroma Scudder 1875b: 90. Type species: *Stenobothrus quadrimaculatus* Thomas, by monotypy.

Prorocorypha Rehn 1911: 301. Type species: *Prorocorypha snowi* Rehn, by original designation. Placed under *Mesopsis* by Jago 1969.

Pseudopomala Morse 1896a: 342. Type species: *Opomala brachyptera* Scudder, by original designation. Placed under *Chloealtis* by Jago 1969.

Psoloessa Scudder 1875a: 512. Type species: *Psoloessa texana* Scudder, subsequent designation by Kirby 1910.

Stirapleura Scudder 1876a: 510. Type species: *Stirapleura decussata,* by monotypy. NEW SYNONYM.

Rhammatocerus Saussure 1861: 317. Type species: *Stenobothrus (Rhammatocerus) viatorius* Saussure, subsequent designation by Rehn 1940.

Pseudostauronotus Brunner de Wattenwyl 1893: 123. No included species. Synonymized by Rehn 1940.

Plectrophorus McNeill 1897b: 251. No type designated; genus included *P. viatorius* and *P. gregarius*. Homonym, replaced by *Plectrotettix* McNeill 1897a: 71.

Plectrotettix McNeill 1897a: 71. Replacement name for *Plectrophorus* McNeill. Synonymized by Rehn 1940.

Scyllina Stål 1873: 94, 112. Type species: *Gomphocerus (Epacromia) peregrans* Stål, subsequent designation by Rehn 1906b: 40. Synonymized by Jago 1971.

Cauratettix Roberts 1937: 349. Type species: *Cauratettix gracilis* Roberts, by original designation. Synonymized by Jago 1971.

Silvitettix Bruner 1904: 55. Type species: *Silvitettix communis* Bruner, by monotypy.

Ochrotettix Bruner 1904: 56. Type species: *Ochrotettix salinus* Bruner, by monotypy. Synonymized by Jago 1971.

Leuconotus Bruner 1904: 57. Type species: *Leuconotus biolleyi* Bruner, by monotypy. Synonymized by Jago 1971.

Oaxacella Hebard 1932: 231. Type species: *Oaxacella whitei* Hebard, by original designation. Synonymized by Jago 1971.

Stenobothrus Fisher 1853. Type species: *Gryllus lineatus* Panzer, subsequent designation by Fisher 1853.

Subgenus *Bruneria* McNeill 1897b. Type species: *Gomphocerus shastanus* Scudder, by original designation. Subgeneric assignment by Jago 1971.

Stethophyma Fisher 1853. Type species: *Gryllus (Locusta) grossus* Linn. 1758, subsequent designation by Roberts 1941b: 13. See also discussion by Jago 1971: 215.

Syrbula Stål 1873: 102. Type species: *Syrbula leucocerca* Stål, by subsequent designation. Synonym of *Syrbula admirabilis* (Uhler).

Xeracris Caudell 1915: 25. Type species: *Heliastus minimus* Scudder, by original designation.

Coniana Caudell 1915: 26. Type species: *Coniana snowi* Caudell, by original designation. Synonymized by Jago 1971.

The Species of Gomphocerinae and Acridinae

This appendix lists in alphabetical order all species of Gomphocerinae and Acridinae (boldface) and their synonyms (italics). Literature references, type localities, and all generic reassignments are included. The present locations of the type are given by the following abbreviations:

AMNH American Museum of Natural History, New York
ANSP Academy of Natural Sciences of Philadelphia, Philadelphia
BM British Museum of Natural History, London
CAS California Academy of Sciences, San Francisco
CNC Canadian National Collection, Ottawa
GM Muséum d'Histoire Naturelle, Geneva
LEM Lyman Entomological Museum, McGill University, Montreal
MM Instituto Español de Entomologia, Madrid
PM Muséum National d'Histoire Naturelle, Paris
SM Naturhistoriska Riksmuseum, Stockholm
TM Museo ed Istituto di Zoologia Sistematica del 'Universita, Torino
UMMZ University of Michigan Museum of Zoology, Ann Arbor
UN University of Nebraska State Museum, Lincoln
USNM United States National Museum, Washington, D.C.
VM Naturhistorisches Museum, Vienna

Acantherus piperatus Scudder and Cockerell 1902: 23. Holotype ♀, La Cueva, Organ Mts., New Mexico, 5,300 ft. ANSP.
Achurum carinatum (Walker)
    Mesops carinatus Walker 1870: 501. Holotype ♂, North America. BM. Transferred to *Radinotatum* by Uvarov 1925: 268. Transferred to *Achurum* by Jago 1969: 282.
    Truxalis brevipennis Thomas 1873b: 58. Holotype ♀, temperate Mexico. Vienna museum. Transferred to *Achurum* by Scudder 1877: 29. Transferred to *Radinotatum* by McNeill 1897b: 200.
    Radinotatum brevipenne peninsulare Rehn and Hebard 1912: 246. Holotype ♂, Homestead, Dade Co., Florida. ANSP. Synonymized under *Achurum carinatum brevipenne* by Jago 1969: 287.

Achurum minimipenne Caudell 1904: 110. Holotype ♀, Brownsville, Texas. Brooklyn Children's Museum.

Achurum sumichrasti (Saussure)

Truxalis sumichrasti Saussure 1861: 313. Holotype ♀, temperate Mexico. VM. Transferred to *Achurum* by McNeill 1897b: 202.

Truxalis acridoides Stål 1873: 52. Holotype ♂, Mexico. SM. Synonymized by Hebard 1922: 91.

Achurum hilliardi Gurney 1959. Holotype ♂, 1 mi S of Warren, Tyler Co., Texas. USNM. NEW SYNONYM.

Acrolophitus hirtipes (Say)

Gryllus hirtipes Say 1825. Type ♀, Colorado, banks of Arkansas R., 150 mi from mountains. Type lost. Transferred to *Acrolophitus* by Thomas 1871a: 278.

Acrolophitus uniformis Bruner 1904: 47. Type ♀, Sterling, Colorado. ANSP. Synonymized by Rehn and Hebard 1906: 363.

Acrolophitus variegatus Bruner 1904: 48. Lectotype ♂, designated by Rehn and Hebard 1912a, Carrizo Springs, Dimmit Co., Texas. ANSP. Synonymized by Jago 1969. The subspecific status was discussed by Hebard (1937) in some unpublished notes.

Acrolophitus nevadensis (Thomas) NEW COMBINATION

Pedioscirtetes nevadensis Thomas 1873c: 295. Holotype ♀, SE Nevada. USNM.

Acrolophitus maculipennis (Scudder) NEW COMBINATION

Acrocara maculipenne Scudder 1890: 437. Lectotype ♂, designated by Jago 1969, Montelovez, Coahuila, Mexico. ANSP. Transferred to *Pedioscirtetes* by Hebard 1926: 50.

Acrolophitus pulchella (Bruner) NEW COMBINATION

Pedioscirtetes pulchella Bruner 1890: 60. Lectotype ♂, designated by Rehn and Hebard 1912a, Birch Creek, Idaho, August. USNM. Treated by White and Nickerson 1951: 239 as a synonym of *Pedioscirtetes nevadensis*.

Aeropedellus arcticus Hebard 1935: 207. Holotype ♀, latitude 60° 20′ N, long. 141° W, Malcolm R., International Boundary, Alaska. USNM.

Aeropedellus clavatus (Thomas)

Gomphocerus clavatus Thomas 1873b: 96. Holotype ♂, Kansas. Specimen probably mislabeled. Alexander 1961 demonstrated that the type locality is Pikes Peak, Colorado. USNM. Transferred to *Aeropedellus* by Hebard 1935: 187.

Gomphocerus carpenterii Thomas 1874: 65. Holotype ♂, near Mountain of the Holy Cross, Colorado, 8,000–10,000 ft. USNM. Synonymized by Hebard 1935: 187.

Gomphocerus clepsydra Scudder 1875b: 344. Holotype ♀, Souris R., Manitoba, long. 99°–102° E. ANSP. Synonymized by Hebard 1935: 187.

Ageneotettix brevipennis (Bruner) NEW COMBINATION

Zapata brevipennis Bruner 1904: 103. Holotype ♀, Lerdo, Durango, Mexico. ANSP.

Zapata bucculenta Rehn 1927: 219. Holotype ♀, Durango (city), Durango, Mexico. USNM. NEW SYNONYM.

Ageneotettix deorum (Scudder)

Chrysochraon deorum Scudder 1876a: 262. Holotype ♂, Garden of the Gods, Colorado. ANSP. Transferred to *Eremnus,* homonym of *Eremnus* Schon, McNeill 1897: 268. Transferred to *Ageneotettix* by McNeill 1897b: 31.

Aulocara scudderi Bruner 1890: 63. Transferred to *Eremnus* by McNeill 1897b: 269. *Eremnus* preoccupied, transferred to *Ageneotettix* by McNeill 1897b. Synonymized by Rehn and Hebard 1906.

Ageneotettix occidentalis Bruner 1905: 109. Lectotype ♂, designated by Rehn and Hebard 1912a, Glenwood Springs, Colorado. ANSP. Synonymized by Rehn and Hebard 1906: 371.

Ageneotettix curtipennis Bruner 1905: 109. Type ♀, Durango, Colorado, August 7, 1899. ANSP. *Ageneotettix deorum curtipennis* Hebard 1935: 284. NEW SYNONYM.

Ageneotettix australis Bruner 1905: 110. Lectotype ♀, designated by Rehn and Hebard 1912a: 113, Phoenix, Arizona. ANSP. Synonymized by Hebard 1935: 283.

Ageneotettix arenosus Hancock, 1906: 255. ♂, ♀ types, Chicago, Illinois. ANSP. Synonymized by Hart 1906.

Ageneotettix salutator (Rehn) NEW COMBINATION

Zapata salutator Rehn 1927: 221. Holotype ♀, Tumamoc Hill, Tucson Mts., Pima Co., Arizona. ANSP.

Amblytropidia elongata Brunner 1904: 68. Lectotype ♀, designated by Rehn and Hebard 1912a, Tepic, Nayarit, Mexico. ANSP.

Amblytropidia bruneri Hebard 1932: 233. Holotype ♂, Medellín, Veracruz, Mexico. ANSP. NEW SYNONYM.

Amblytropidia mysteca (Saussure)

Stenobothrus mystecus Saussure 1861: 317. Holotype ♀, Orizaba, Veracruz, Mexico. GM. Transferred to *Amblytropidia* by Rehn 1902a: 9.

Stenobothrus occidentalis Saussure 1861: 317. Holotype ♂, Tennesse. GM. Transferred to *Amblytropidia* by McNeill 1897: 226. NEW SYNONYM.

Stenobothrus subconspersus Walker 1870: 755. Holotype ♂, St. John's Bluff, eastern Florida. BM. Synonymized by Scudder 1899b under *A. occidentalis.*

Amblytropidia subhyalina Scudder 1875a: 511. Holotype ♂, Dallas, Texas. ANSP. Synonymized by Bruner 1904: 68.

Amblytropidia auriventris McNeill 1897b: 227. Holotype ♂, Orizaba, Veracruz, Mexico. ANSP. Synonymized by Hebard 1925a: 271.

Amblytropidia costaricensis Bruner 1904: 66. Lectotype ♂, designated by Rehn and Hebard 1912, San José, Costa Rica. ANSP. NEW SYNONYM.

Amblytropidia ingenita Bruner 1904: 67. ♂, Orizaba, Veracruz; ♂, ♀, Chilpancingo, Guerrero; ♂, ♀ Cuernavaca, Morelos, Mexico. Type series lost? Synonymized by Hebard 1925.

Amblytropidia trinitatis Bruner 1904: 65. ♂, ♀♀, Trinidad. UN.

Amblytropidia magna Bruner 1904: 63. Holotype ♀, Panzós, Alta Verapaz, Guatemala. Type lost. NEW SYNONYM.

Amblytropidia insignis Hebard 1923: 198. Holotype ♂, Gatun, Panama. ANSP. NEW SYNONYM.

Amblytropidia pulchella Hebard 1932: 234. Holotype ♂, Atoyac, Veracruz, Mexico. ANSP. NEW SYNONYM.

Amphitornus coloradus (Thomas)

Stenobothrus coloradus Thomas 1873b: 82. Lectotype ♀, designated by Hebard 1927: 3, Colorado. USNM. Transferred to *Amphitornus* by Hebard 1925b: 54.

Amphitornus ornatus McNeill 1897b: 225. Type series, Los Angeles, California. USNM. Synonymized by Hebard 1937: 357 in a footnote.

Akentetus unicolor McNeill 1897b: 225. Holotype ♂, Colorado. Type lost. Synonymized by Hebard 1925b: 54.

Amphitornus nanus Rehn and Hebard 1908: 376. Holotype ♂, rim of Coconino Plateau, Grand Canyon, Arizona. ANSP. Synonym of *A. ornatus,* Hebard 1935: 280. Synonymized by Hebard 1937: 356.

Amphitornus coloradus saltator Hebard 1937: 357. Holotype ♂, Shell Creek Range, E of road summit, White Pine Co., Nevada. ANSP.

Amphitornus durangus Otte 1979b. Holotype ♂, 14.1 mi ENE of Llano Grande on Highway 40, Durango, Mexico. UMMZ.

Aulocara brevipenne Bruner 1905: 111. Lectotype ♂, designated by Rehn and Hebard 1912a: 113, Comancho, Zacatecas, Mexico. ANSP. The specimen chosen by Rehn and Hebard as lectotype was originally labeled a paratype of *Zapata brevipennis* Bruner; the holotype was evidently lost.

Aulocara elliotti (Thomas)

Stauronotus elliotti Thomas 1870: 82. Lectotype ♀, designated by Hebard 1927: 4, eastern Colorado. USNM. Transferred to *Aulocara* by Bruner 1885: 10.

Aulocara caeruleipes Scudder 1876a: 266. Type ♂♂, Garden of the Gods, Colorado. ANSP. Synonymized by Bruner 1905: 111.

Aulocara decens Scudder 1876a: 267. Holotype ♂, Lake Point, Salt Lake, Utah. ANSP. Synonymized by Bruner 1905: 111.

Oedocara strangulatum Scudder 1876a: 289. Holotype ♀, S. Col. (= southern Colorado?). ANSP. Synonymized by Bruner 1905: 111.

Aulocara parallelum Scudder 1899b: 57. Lectotype ♂, designated by Rehn and Hebard 1912, Salt Lake Valley, Utah. ANSP. NEW SYNONYM.

Aulocara femoratum Scudder 1899b: 55. Lectotype ♂, designated by Rehn and Hebard 1912a, Lakin, Kansas. ANSP. Transferred to *Drepan-*

opterna by Rehn 1927: 226. Transferred to *Aulocara* by Jago 1971: 262.

Boopedon auriventris McNeill 1899: 54. Holotype ♂, Camp 22, Sulphur Springs, Arkansas. ANSP.

Boopedon savannarum Bruner 1904: 97. Lectotype ♂, designated by Rehn and Hebard 1912a, West Point, Nebraska. ANSP. Synonymized by Hebard 1934: 29.

Boopedon dampfi (Hebard)

Morseiella dampfi Hebard 1932: 241. Holotype ♂, San Geronimo, Oaxaca, Mexico. ANSP. Transferred to *Boopedon* by Jago 1971: 263.

Boopedon diabolicum Bruner 1904: 98. Lectotype ♀, designated by Rehn and Hebard 1912a, Tepic, Nayarit, Mexico. ANSP. Transferred to *Morseiella* by Hebard 1932: 239. Transferred to *Boopedon* by Jago 1971: 263.

Boopedon empelios Otte 1979b. Holotype ♂, 16 mi E of Navajoa on Alamos highway, Sonora, Mexico, 800 ft. ANSP.

Boopedon flaviventris (Bruner)

Boopedon flaviventris Bruner 1904: 98. Lectotype ♀, designated by Rehn and Hebard 1912a, Tepic, Nayarit, Mexico. ANSP. Transferred to *Morseiella* by Hebard 1932: 239. Transferred to *Boopedon* by Jago 1971: 263.

Boopedon gracile Rehn 1904: 520. Holotype ♂, Altamira, Tamaulipas, Mexico. ANSP.

Boopedon nubilum (Say)

Gryllus nubilus Say 1825a: 308. Type ♂, eastern Colorado on Arkansas R. (Long Expedition). Type lost. Transferred to *Boopedon* by Thomas 1871a: 272.

Boopedon nigrum Thomas 1870: 83. Types from New Mexico and Canon City, southern Colorado, near the mountains. Types lost. Synonymized by Bruner 1904: 96.

Boopedon flavofasciatum Thomas 1870: 84. Lectotype ♀ southern Colorado and New Mexico near mountains. USNM. Synonymized by Caudell 1915: 29.

Boopedon fuscum Bruner 1904: 96. Lectotype ♂, designated by Rehn and Hebard 1912, Nogales, Arizona. ANSP. Synonymized by Hebard 1935: 285.

Boopedon hoagi Rehn 1904: 519. Holotype ♂, LaJoya, San Luis Potosí, Mexico. ANSP. NEW SYNONYM.

Boopedon nubilum maculatum Caudell 1915: 29. Holotype ♂, Victoria, Texas. USNM. Assigned specific status by Hebard 1931. Synonymized by Hebard 1937.

Boopedon rufipes (Hebard)

Morseiella rufipes Hebard 1932: 240. Holotype ♂, Morelos, Mexico. ANSP. Transferred to *Boopedon* by Jago 1971: 263.

Bootettix argentatus Bruner

Bootettix argentatus Bruner 1890: 58. Lectotype ♂, designated by Rehn and Hebard 1912a. Lerdo, Durango, Mexico. ANSP.

Gymnes punctatus Scudder 1890: 440. Holotype ♀, California. ANSP. Synonymized by Bruner 1904 under *B. argentatus*. Supspecies of *B. argentatus*, Jago 1969: 240.

Bootettix joerni Otte 1979b. Holotype ♂, 36 mi NE of San Pedro de las Colonias, on Rt 30 NE of Torreón, Coahuila, Mexico. ANSP.

Chiapacris eximius Otte 1979b. Holotype ♂, Cerro Tancitaro, 2 km N of Apo, Michoacán, Mexico. UMMZ.

Chiapacris nayaritus Otte 1979b. Holotype ♂, Campostella, Nayarit, Mexico. UMMZ.

Chiapacris velox Otte 1979b. Holotype ♂, 3 km N of Pueblo Nuevo, Chiapas, Mexico. ANSP.

Chloealtis abdominalis (Thomas)

Chrysochraon abdominalis Thomas 1873b: 74. Holotype ♀, Montana. USNM. Transferred to *Chloealtis* by McNeill 1897b: 229. Transferred to *Neopodismopsis* by Bei-Bienko 1932: 56. Transferred to *Chloealtis* by Jago 1969: 294.

Acentetus carinatus Scudder 1899b: 45. Holotype ♂, Florissant, Colorado. ANSP. Emendation of *Akentetus*. Synonym of *Amphitornus coloradus* Hebard 1925b: 54. NEW SYNONYM.

Chloealtis aspasma Rehn and Hebard 1919: 82. Holotype ♀, Jackson Co., Siskiyou Mts., Oregon, 5,000–5,800 ft. ANSP. Transferred to *Napaia* by Rehn 1928. Transferred to *Chloealtis* by Jago 1969: 293.

Chloealtis conspersa (Harris)

Locusta (*Chloealtis*) *conspersa* Harris 1841: 149. Type series lost, Morse 1921: 211. Neotype ♂, Jago 1969. ANSP.

Chloealtis dianae (Gurney, Strohecker, and Helfer)

Napaia dianae Gurney, Strohecker, and Helfer 1964: 125. Holotype ♂, Mill Creek, Mendocino Co., California. USNM. Transferred to *Chloealtis* by Jago 1969.

Chloealtis gracilis (McNeill)

Napaia gracilis McNeill 1897b: 213. Holotype ♂, Los Angeles Co., California. USNM. Transferred to *Chloealtis* by Jago 1969.

Oenomus altus Scudder 1899b: 47. Lectotype ♂, designated by Rehn and Hebard 1912, Mt. Wilson, Calfornia, 2,400 ft. ANSP. Synonymized by Bruner 1904: 90.

Chorthippus curtipennis (Harris)

Locusta curtipennis Harris 1835: 576. Neotype ♂, selected by Vickery 1964, Waltham, Massachusetts. UMMZ. Transferred to *Stenobothrus* by Scudder 1862: 286. Transferred to *Chorthippus* by Rehn 1902: 316. Transferred to *Stauroderus* by Rehn and Hebard 1906: 369. Synonym of *Chorthippus longicornis* Latreille, Hebard 1936. Name restored by Vickery 1964.

Stenobothrus longipennis Scudder 1862: 457. Holotype ♂, Cambridge, Massachusetts. ANSP. Synonym of *Chorthippus longicornis* Latreille, Hebard 1936. Synonym of *C. curtipennis*, Vickery 1964.

Stenobothrus coloradensis McNeill 1897b: 262. Holotype ♀, Fort Collins, Colorado. Type lost. Synonym of *C. longicornis* Latreille, Hebard 1936. Synonym of *C. curtipennis*, Vickery 1964.

Stenobothrus oregonensis Scudder 1899b: 50. Lectotype ♂, designated by Rehn and Hebard 1912a, Divide, Oregon. Type lost, not among Scudder types at ANSP. Synonym of *C. longicornis* Latreille, Hebard 1936. Synonym of *C. curtipennis*, Vickery 1964.

Stenobothrus acutus Morse 1903: 115. Lectotype ♂, designated by Morse and Hebard 1915, W of Carson City, Ormsby Co., Nevada. ANSP. Synonym of *C. longicornis*, Hebard 1936. Synonym of *C. curtipennis*, Vickery 1964.

Chorthippus curtipennis californicus Vickery 1967: 113. Holotype ♂, Mendocino, California. LEM.

Chrysochraon petraea (Gurney, Strohecker, and Helfer)

Barracris petraea Gurney, Strohecker, and Helfer 1964: 122. Holotype ♂, Meadow Lake, Gilmore, Lemhi Co., Idaho, July 27, 1961. CAS. Transferred to *Chrysochraon* by Jago 1971: 251.

Cibolacris crypticus (Vickery) NEW COMBINATION

Astehelius crypticus Vickery 1969a: 1224. Holotype ♂, 40 mi SW of Ciudad Obregón, near San José Beach, Sonora, Mexico. CNC.

Cibolacris parviceps (Walker)

Oedipoda parviceps Walker 1870: 732. Holotype ♂, west coast of North America. BM.

Thrincus aridus Bruner 1890: 78. Lectotype ♂, designated by Rehn and Hebard 1909: 461. Synonymized by Hebard 1937: 368.

Thrincus californicus Thomas 1874: 66. Southern California. Type lost? Transferred to *Heliastus* by Scudder 1897: 75. Synonymized by Hebard 1937: 374.

Cibolacris samalayucae Tinkham 1961: 30. Holotype ♂, Samalayuca Dunes at road station La Noria, 33 mi S of El Paso, Texas, in Chihuahua, Mexico. Tinkham Collection. Indio, California.

Cibolacris weissmani

Cibolacris weissmani new species. Holotype ♂, 11.2 km along dirt road to Scammons Lagoon (Lago Ojo de Liebre) off Mexico Highway 1, Baja California Sur, Mexico, July 11, 1978 (Weissman and Lightfoot) California Academy of Sciences.

Compsacrella poecila Rehn and Hebard 1938: 209. Holotype ♂, 12.5 km S of Pinar del Rio, Pinar del Rio Province, Cuba. AMNH.

Cordillacris occipitalis (Thomas)

Stenobothrus occipitalis Thomas 1873b: 81. Lectotype ♀, designated by Hebard 1925b: 55, Colorado. USNM. Transferred to *Cordillacris* by Rehn 1901.

Ochrilidea cinerea Bruner 1890: 52. Lectotype ♂, designated by Rehn and Hebard 1912a, Fort McKinley, Wyoming. ANSP. Synonymized by Hebard 1928: 227 as subspecies.

Cordillacris affinis Morse 1903: 115. Lectotype ♂, designated by Morse and Hebard 1915: 99, lower edge of pine zone, W of Carson City,

Ormsby Co., Nevada, 1,700–2,000 m. ANSP. Synonymized by Jago 1969: 278.

Cordillacris grinnelli Rehn and Hebard 1910: 425. Holotype ♀, south fork of Santa Ana R., San Bernardino Mts., California. ANSP. Synonymized by Jago 1969: 278.

Cordillacris crenulata (Bruner)

Ochrilidia(?) *crenulata* Bruner 1890: 51. Lectotype ♂, designated by Rehn and Hebard 1912, Fort Robinson, Nebraska. ANSP.

Cordillacris pima Rehn 1907: 69. Holotype ♀, Baboquivari Mts., Pima Co., Arizona. University of Kansas collection. Synonymized by Hebard 1935: 281, who considered it a subspecies of *C. crenulata*.

Cordillacris apache Rehn and Hebard 1909: 139. Lectotype ♀, designated by Rehn and Hebard 1912a: 105, Silver City, Grant Co., New Mexico. ANSP.

Dichromorpha elegans (Morse)

Clinocephalus elegans Morse 1896: 402. Lectotype ♂, designated by Morse and Hebard 1915, Ravenswood, Long Island, New York. ANSP. Transferred to *Dichromorpha* by Otte 1979a.

Clinocephalus pulcher Rehn and Hebard 1905: 36. Lectotype ♂, designated by Rehn and Hebard 1912, Miami, Florida. ANSP. Synonym of *C. elegans,* Rehn and Hebard 1916: 172.

Dichromorpha prominula (Bruner)

Orphulella prominula Bruner 1904: 82. Lectotype ♂, here designated, Mazatlán, Mexico. ANSP. Transferred to *Dichromorpha* by Hebard 1932: 238.

Dichromorpha longipennis Bruner 1904: 87. Lectotype ♀, designated by Rehn and Hebard 1912a, Tepic, Nayarit, Mexico. ANSP. Synonymized by Hebard 1932: 239.

Dichromorpha viridis (Scudder)

Chloealtis viridis Scudder 1862: 455. Holotype ♂, Norton, Connecticut. ANSP. Transferred to *Stenobothrus* by Walker 1870: 755. Transferred to *Chrysochraon* by Thomas 1873. Transferred to *Tryxalis* by Thomas 1876. Transferred to *Dichromorpha* by Morse 1896: 383.

Orphulella robusta Bruner 1904: 83. Lectotype ♂, here designated, Amula, Guerrero, Mexico. ANSP. Transferred to *Dichromorpha* by Hebard 1932: 238. Synonymized by Otte 1979a.

Dichromorpha mexicana Bruner 1904: 87. Lectotype ♀, designated by Rehn and Hebard 1912a, Tepic, Nayarit, Mexico. ANSP. Synonym of *Dichromorpha robusta,* Hebard 1932: 238.

Eritettix abortivus (Bruner)

Eritettix abortivus Bruner 1890: 56. Lectotype ♂, designated by Rehn and Hebard 1912, Washington Co., Texas. ANSP. Transferred to *Mesochloa* by Scudder 1898: 234. Transferred to *Eritettix* by Jago 1971: 246.

Mesochloa unicolor Hart 1906: 157. 8 ♂♂, 11 ♀♀, Texas Agricultural

College, Brazos Co., Texas. Illinois State Laboratory of Natural History collection. Synonymized by Hebard 1936.

Eritettix obscurus (Scudder) NEW COMBINATION

Chrysochraon obscurus Scudder 1878b: 88. Holotype ♂, Fort Reed, Florida. ANSP. Type partially destroyed. Transferred to *Pedeticum,* McNeill 1897b: 216. Homonym of *Pedeticus* Laporte, transferred to *Macneillia* by Scudder 1898: 235. Transferred to *Amphitornus* by Jago 1971: 270.

Eritettix sylvestrus Blatchley 1902: 219. Type lost. Synonymized by Rehn and Hebard 1912a.

Eritettix simplex (Scudder)

Gomphocerus simplex Scudder 1869a: 305. Holotype ♂, Delaware. ANSP.

Gomphocerus carinatus Scudder 1875a: 511. 2 ♀ types, Maryland. ANSP. Synonymized by Rehn and Hebard 1910: 626.

Gomphocerus virgatus Scudder 1875a: 511. 22 ♀ types, Dallas, Texas. ANSP. Transferred to *Eritettix* by Bruner 1889: 57. Synonymized by Hebard 1931: 140.

Stenobothrus tricarinatus Thomas 1873b: 84. Holotype ♀, Wyoming. Type lost, Hebard 1927: 2. Transferred to *Fritettix* by McNcill 1897: 218. Placed as a subspecies of *E. simplex* by Hebard 1936: 29.

Gomphocerus navicula Scudder 1876b: 286. Type ♂, southern Colorado. ANSP. Transferred to *Eritettix* by McNeill 1897b: 220. Synonym of *E. tricarinatus,* Hebard 1929: 326. Synonymized by Hebard 1936: 29.

Eritettix variabilis Bruner 1890: 56. Lectotype ♂, designated by Rehn and Hebard, Silver City, New Mexico. USNM. NEW SYNONYM.

Eritettix vernalis Bruner 1893: 22. *Nomen nudum.*

Eupedetes carinatus Scudder 1902. Holotype ♂, La Trementina, New Mexico. ANSP. Synonymized by Rehn 1907.

Eritettix simplex dorsalis Blatchley 1920: 212. *Nomen nudum.*

Eritettix brachypterus Bruner 1904: 53. Type ♀, Durango, Mexico, 8,100 ft. Type lost. *Nomen dubium.*

Esselenia vanduzeei Hebard

Esselenia vanduzeei Hebard 1920: 73. Holotype ♀, Bryson, Monterey Co., California. CAS.

Esselenia vanduzeei violae Rentz 1966: 5. Holotype ♂, Russelmann Park, NE slope of Mt. Diablo, Contra Costa Co., California. CAS.

Eupnigodes megacephala (McNeill)

Pnigodes megacephala McNeill 1897b: 267. Type series, Yuba and Butte Cos., California. USNM. *Pnigodes* preoccupied, transferred to *Eupnigodes* by McNeill, 1897b: 71.

Eupnigodes sierranus Rehn and Hebard

Ageneotettix sierranus Rehn and Hebard 1910: 429. Lectotype ♂, designated by Rehn and Hebard 1912a, Summit House, Madera Co., California. ANSP. Transferred to *Eupnigodes* by Rehn 1942: 174.

Heliaula rufa (Scudder)

Aulocara rufum Scudder 1899b: 55. Lectotype ♂, designated by Rehn and Hebard 1912a, Pueblo, Colorado. ANSP. Transferred to *Heliaula* by Caudell 1915: 27.

Horesidotes cinereus Scudder 1899b: 49. Lectotype ♂, designated by Rehn and Hebard 1912a: 91, Palm Canyon, Palm Springs, California. ANSP.

Horesidotes papagensis Rehn and Hebard 1908: 379. Holotype ♀, Sonora Road Canyon, Tucson Mts., Pima Co., Arizona, 3,000 ft. ANSP. Synonymized by Hebard 1935: 282.

Horesidotes cinereus saltator Hebard 1931: 116. Holotype ♂. Tia Juana, California. ANSP.

Horesidotes deiradonotus (Jago) NEW COMBINATION

Silvitettix deiradonotus Jago 1971: 276. Holotype ♂, 49 mi E junction of highways 15 and 40, on 40, Durango, Mexico. ANSP.

Leurohippus stoneri (Caudell)

Amblytropidia stoneri Caudell 1922: 32. Holotype ♂, Antigua. USNM. Transferred to *Caribacris* by Rehn and Hebard 1938: 202. Transferred to *Leurohippus* by Jago 1971: 239.

Ligurotettix coquilletti McNeill 1897b: 258. Lectotype ♂, designated by Rehn 1923: 75, Los Angeles Co., California. USNM.

Ligurotettix kunzei Caudell 1903: 162. Holotype ♀, Phoenix, Arizona. USNM. Synonymized by Rehn 1923: 82.

Ligurotettix coquilletti cantator Rehn 1923: 64. Holotype ♂, Mason, Lyon Co., Nevada. ANSP.

Ligurotettix planum (Bruner)

Goniatron planum Bruner 1905: 113. Lectotype ♂, designated by Rehn 1923: 47, Comancho, Zacatecas, Mexico. ANSP. Transferred to *Ligurotettix* by Jago 1971: 282.

Melanotettix dibelonius Bruner 1904: 91. Holotype ♂, Acaguizotla, Guerrero, Mexico.

Mermiria bivittata (Serville)

Opsomala bivittata Serville 1839: 589. Type lost, neotype ♂ selected by Jago 1969, Lane, South Carolina. ANSP.

Mermiria maculipennis Bruner 1904: 54. Lectotype ♀, designated by Rehn and Hebard 1912a, San Antonio, Texas. ANSP. *M. bivittata maculipennis,* Jago 1969.

Mermiria maculipennis macclungi Rehn 1919: 111. Holotype ♂, Forsyth, Rosebud Co., Montana. ANSP. *M. bivittata maculipennis,* Jago 1969.

Mermiria intertexta Scudder 1899b: 42. Lectotype ♂, designated by Rehn and Hebard 1912a, Georgia. ANSP.

Mermiria picta (Walker)

Opomala picta Walker 1870: 516. Holotype ♂, locality unknown. BM. Transferred to *Mermiria* by Uvarov 1925.

Opomala neomexicana Thomas 1870: 77. Holotype ♀, NE New Mexico. USNM. *M. picta neomexicana,* Jago 1969: 275.

Mermiria belfragii Stål, 1873: 102. Holotype ♀, Texas. SM. Synonymized by Jago 1969: 274.

Mermiria alacris Scudder 1878a: 30. Lectotype ♂, designated by Jago 1969, Georgia. ANSP. Synonymized by Uvarov 1925: 268.

Mermiria rostrata McNeill 1897b: 207. ♂♂, ♀, Oklahoma, Indian Territory. Type lost. Synonymized by Uvarov 1925.

Mermiria vigilans Scudder 1899b: 43. Lectotype ♂, designated by Rehn and Hebard 1912a, Smithville, North Carolina. ANSP. Synonymized by Uvarov 1925.

Mermiria texana Bruner 1890: 53. Lectotype ♂, designated by Rehn and Hebard 1912, El Paso, Texas. ANSP.

Papagoa arizonensis Bruner 1904: 42. Holotype ♂, Arizona or northern Mexico. ANSP. Synonymized by Rehn 1919: 65.

Metaleptea brevicornis (Johannson)

Gryllus brevicornis Johannson 1763: 398. North America. Location of type unknown. Giglio-Tos 1897: 22 was the first to combine the name *brevicornis* with *Metaleptea*. Transferred to *Truxalis* by Kirby 1910: 103. Transferred to *Metaleptea* by Gurney 1940: 92.

Acrydium ensicornum DeGeer 1773: 499. Holotype ♀, Pennsylvania. SM. Synonymized by Kirby 1910: 103.

Truxalis notochloris Palisot de Beauvoir 1805: 80. Type ♀, Santo Domingo (Rehn 1944 believes the locality to be in error). Location of type not known. Synonymized by Kirby 1910: 103.

Opsomala punctipennis Serville 1839: 590. Type ♀, North America. According to Dr. Carbonell the type is lost. Synonymized by Kirby 1910: 104.

Oxycoryphus burkhartianus Saussure 1861: 314. Holotype ♀, Mexico. GM. Synonymized by Kirby 1910: 104.

Opomala stenobothroides Walker 1871: 52. Holotype ♂, Chontales, Nicaragua. BM. Synonymized by Kirby 1904: 104.

Opeia obscura (Thomas)

Oxycoryphus obscurus Thomas 1872: 466. Lectotype ♀, designated by Hebard 1925b: 53, Fort Fetterman, Wyoming. USNM. Transferred to *Opeia* by McNeill 1897b: 214.

Opeia testacea Scudder 1899b: 46. Lectotype ♂, designated by Rehn and Hebard 1912a, Lancaster, California. ANSP.

Opeia palmeri Bruner 1904: 61. Lectotype ♂, designated by Rehn and Hebard 1912a, Sierra de San Miguelito, Mexico. ANSP.

Opeia mexicana Bruner 1904: 60. Holotype ♀, Tlalpam, near Mexico City. ANSP. NEW SYNONYM.

Opeia lineata Bruner 1904: 61. Holotype ♀, Sierra de San Miguelito, Mexico. Type lost, should be in Bruner collection but is not in ANSP. *O. mexicana,* Hebard 1937: 354 in a footnote.

Opeia imperfecta Bruner 1904: 59. Lectotype ♀, designated by Rehn and Hebard 1912a, Jimulco, Mexico. ANSP.

Opeia pallida Bruner 1904: 60. Holotype ♂, Montelovez, Coahuila,

Mexico. Type lost, should be in Bruner collection but is not in ANSP.

Opeia atascosa Hebard

Opeia atascosa Hebard 1937: 354. Holotype ♂, SW slopes of Atascosa Peak near Bear Valley, Pajarito Mts., Arizona. ANSP.

Orphula azteca (Saussure)

Oxycoryphus aztecus Saussure 1861: 315. Holotype ♂, Cordova, Mexico. GM.

Oxycoryphus mexicanus Saussure 1861: 314. Holotype ♀, Tescutlan, Mexico. GM. Transferred to *Orphulella* by Bruner 1904: 80. Synonymized by Hebard 1923: 207 under *Orphulella punctata*. NEW SYNONYM.

Orphula guatemalae Bruner 1906: 10. Holotype ♂, Santa Lucia, Guatemala. ANSP. Synonymized by Hebard 1932: 236.

Orphula meridionalis Bruner 1904: 73. Holotype ♀, Pozo Azul, Costa Rica. ANSP. Synonymized by Hebard 1932: 236.

Orphulella neglecta Rehn 1900: 94. Holotype ♂, Orizaba, Veracruz, Mexico. ANSP. Synonymized by Hebard 1932: 236.

Sisantum notochloris Bruner 1904: 69. Lectotype ♀, designated by Rehn and Hebard 1912, Medellín, Veracruz, Mexico. ANSP. NEW SYNONYM.

Orphula vitripenne (Bruner) NEW COMBINATION

Thyriptilon vitripenne Bruner 1904: 69. Lectotype ♂, designated by Rehn and Hebard 1912, Vallodolid, Yucatán, Mexico. ANSP.

Orphulella aculeata Rehn

Orphulella aculeata Rehn 1900: 92. Holotype ♂, Cuernavaca, Morelos, Mexico. ANSP.

Orphulella brachyptera Rehn and Hebard 1938: 206. Holotype ♂, 12.5 km S of Pinar del Rio, Cuba. AMNH.

Orphulella concinnula (Walker)

Stenobothrus concinnulus Walker 1870: 759. Holotype ♂, Pará, Brazil. BM. Transferred to *Orphulina* by Kirby 1910. Transferred to *Orphulella* by Uvarov 1925: 265.

Stenobothrus rugulosus Walker 1870: 760. Holotype ♂, Santarém, Brazil. BM. Synonymized by Otte 1979.

Linoceratium boucardi Bruner 1904: 84. Lectotype ♂, designated by Rehn and Hebard 1912a, Panama. ANSP. Transferred to *Orphulella* by Rehn 1916: 217. Synonymized by Hebard 1933: 44.

Orphulella chipmani Bruner 1906: 149. Holotype ♂, Trinidad. UN. Synonym of *O. boucardi*, Rehn 1916: 277. Synonymized by Hebard 1933: 44.

Linoceratium australe Bruner 1911: 20. Lectotype ♀, designated by Rehn and Hebard 1912a, Corumbá, Brazil. ANSP. Synonym of *Orphulella boucardi*, Rehn 1916: 277. Synonymized by Hebard 1933: 44.

Orphulella peruna Bruner 1911: 16. New name by Bruner for *Zono-*

cerus bilineatus Scudder because the name *bilineatus* was already used for *Stenobothrus bilineatus,* a junior synonym of *Orphulella speciosa.* Holotype ♀, Peruvian Andes. ANSP. Synonym of *O. boucardi,* Hebard 1923: 204. Synonymized by Hebard 1933: 44.

Orphulella decisa (Walker)
   Stenobothrus decisus Walker 1870: 757. Holotype ♂, Santo Domingo, Dominican Republic. BM. Transferred to *Orphulella* by Kirby 1910: 122.

Orphulella losamatensis Caudell 1909: 113. New name given to *Orphulella walkeri* Bruner 1906: 11, a homonym of *Orphulella walkeri* Bruner 1904: 78. Holotype ♂, Los Amates, Guatemala. ANSP.
   Orphulella melanopleura Hebard 1923: 205. Holotype ♂, Andagoya, Antioquia, Colombia. ANSP. Synonymized by Otte 1979a.

Orphulella nesicos Otte 1979a. Holotype ♂, La Vega, 5 km E of Constanza, Dominican Republic. ANSP.

Orphulella orizabae (McNeill)
   Orphulella orizabae McNeill 1897b: 243. Neotype ♂, designated by Otte 1979a, Mexico City. ANSP. Synonymized by Hebard 1932: 237 under *O. tolteca.*
   Orphulella viridescens Scudder 1899a: 187. Type lost, described from Mt. Alvares, Mexico. Synonymized by Hebard 1932: 237 under *O. tolteca.* Since all specimens which Hebard identified as *O. tolteca* actually belong to *O. orizabae,* it is assumed that *O. viridiscens* is a synonym of *O. orizabae.*

Orphulella pelidna (Burmeister)
   Gomphocerus pelidnus Burmeister 1838: 650. Pennsylvania. Location of type not known.
   Stenobothrus maculipennis Scudder 1862: 458. Lectotype ♀, Massachusetts. ANSP. Designation and synonymy by Gurney 1940.
   Stenobothrus propinquans Scudder 1862: 461. Lectotype ♀, Connecticut or Minnesota. ANSP. Designation and synonymy by Gurney 1940.
   Orphulella pratorum Scudder 1899a: 186. Lectotype ♂, Dallas, Texas. ANSP. Designation and synonymy by Gurney 1940.
   Orphulella desereta Scudder 1899a: 184. Lectotype ♂, Salt Lake Valley, Utah. ANSP. Designation and synonymy by Gurney 1940.
   Orphulella salina Scudder 1899a: 185. Lectotype ♂, White R., Colorado. ANSP. Designation and synonymy of Gurney 1940, under *desereta.*
   Orphulella compta Scudder 1899a: 180. Lectotype ♂, designated by Gurney 1940, Palm Springs, California. ANSP. Synonymized by Otte 1979.
   Orphulella affinis Scudder 1899a: 183. Lectotype ♂, designated by Gurney 1940, Palm Springs, California. ANSP. Synonym of *O. compta,* Gurney 1940.
   Orphulella graminea Bruner 1904: 78. Lectotype ♂, designated by

Rehn and Hebard 1912, Phoenix, Arizona. ANSP. Synonym of *O. compta,* Gurney 1940.

Stenobothrus olivaceus Morse 1893: 477. Lectotype ♀, designated by Morse and Hebard 1915, Connecticut. ANSP. Synonymized by Otte 1979.

Orphulella halophila Rehn and Hebard 1916: 166. Holotype ♀, Key West, Monroe Co., Florida. ANSP. Synonym of *O. olivacea,* Gurney 1940.

Orphulella pernix Otte 1979. Holotype ♂, 8.9 mi S of La Cruz on Pan American Highway, Guanacaste Province, Costa Rica. UMMZ.

Orphulella punctata (DeGeer)

Acrydium punctatum DeGeer 1773: 503. Lectotype ♀, designated by Otte 1979. The specimen bears no locality label, only a label "R." AM.

Oxycoryphus totonacus Saussure 1861: 315. Mexico. Location of type unknown, not at GM. Transferred to *Orphulella* by Kirby 1910: 122. Synonymized by Hebard 1923: 208.

Stenobothrus tepanecus Saussure 1861: 319. Holotype ♀, no locality given on label. GM. Synonymized by Otte 1979a.

Oxycoryphus zapotecus Saussure 1861: 316. Holotype ♀, Cordova, Mexico. GM. Synonymized by Otte 1979a.

Stenobothrus mexicanus Walker 1870: 756. Holotype ♂, Oaxaca, Mexico. BM. Synonymized by Bruner 1911: 14.

Stenobothrus viridissimus Walker 1870: 761. Holotype ♀, Honduras. BM. Synonymized by Otte 1979a.

Stenobothrus arctatus Walker 1870: 761. Holotype ♂, no locality given on specimen, but Honduras given as type locality in publication. BM. Synonymized by Otte 1979a.

Stenobothrus expandens Walker 1870: 758. Holotype ♀, Santarém, Brazil. BM. Synonymized by Otte 1979a.

Stenobothrus gratiosus Walker 1870: 758. Holotype ♂, Santarém, Brazil. BM. Synonymized by Otte 1979a.

Stenobothrus costalis Walker 1870: 759. Holotype ♂, Brazil. BM. Synonymized by Otte 1979a.

Truxalis intricata Stål 1873: 106. Holotype ♂, Buenos Aires, Argentina. SM. Transferred to *Orphulella* by Giglio-Tos 1894: 12. Synonymized by Rehn 1906b: 38.

Orphulella gracilis Giglio-Tos 1894: 11. Lectotype ♂, designated by Otte 1979. TM. Synonymized by Otte 1979a.

Orphulella elegans Giglio-Tos 1894: 12. Lectotype ♂, designated by Otte 1979, Villarica, Paraguay. TM. Synonymized by Otte 1979a.

Orphulella costaricensis Bruner 1904: 82. Lectotype ♂, designated by Rehn and Hebard 1912a, San José, Costa Rica. ANSP. Hebard 1923: 207, synonym.

Orphulella meridionalis Bruner 1904: 81. Holotype ♂, Costa Rica. ANSP. Synonymized by Hebard 1923: 207.

Orphulella insularis Bruner 1906: 150. Type series from Trinidad. UN.
Synonymized by Otte 1979a.

Orphulella grossa Bruner 1911: 18. Holotype ♀, Pará, Brazil, April.
ANSP. Synonymized by Otte 1979a.

Orphulella interrupta Bruner 1911: 12. Holotype ♀, Chapada, Brazil.
ANSP. Synonymized by Otte 1979a.

Orphulella compacta Bruner 1911: 19. Holotype ♀, Rio de Janeiro,
Brazil. ANSP. Synonymized by Otte 1979a.

Orphulella quiroga Otte 1979a. Holotype ♂, between Zacapu and Quiroga,
Michoacán, Mexico. ANSP.

Orphulella scudderi (Bolivar)

Orphula scudderi Bolivar 1888: 142. Types, Cuba. MM. Transferred to
Parachloebata by Rehn and Hebard 1938: 203. Transferred to *Orphulella* by Otte 1979.

Parachloebata pratensis Bruner 1904: 84. Holotype ♀, Cuba. Location
of type unknown. Synonymized by Rehn and Hebard 1938: 203.

Orphulella speciosa (Scudder)

Stenobothrus speciosus Scudder 1862: 458. Holotype ♂, St. Paul, Minnesota. ANSP. Transferred to *Orphulella* by Scudder 1899a:
183.

Stenobothrus aequalis Scudder 1862: 459. Lectotype ♀, Sanborn, Massachusetts. ANSP. Designation and synonymy by Gurney 1940.

Stenobothrus bilineatus Scudder 1862: 460. Lectotype ♂, Massachusetts. ANSP. Designation and synonymy by Gurney 1940.

Stenobothrus gracilis Scudder 1872: 250. Holotype ♂, Platte, Nebraska. ANSP. Synonymized by Gurney 1940.

Orphula decora McNeill 1897b: 239. Lectotype ♂, designated by
Hebard 1929, Dallas, Texas. ANSP. Synonymized by Gurney
1940.

Orphulella obliquata Scudder 1899a: 181. Lectotype ♂, designated by
Hebard 1929, Dallas, Texas. ANSP. Synonymized by Gurney
1940.

Orphulella picturata Scudder 1899a: 182. Lectotype ♂, designated by
Hebard 1929, Dallas, Texas. ANSP. Synonymized by Gurney
1940.

Orphulella tolteca (Saussure)

Oxycoryphus toltecus Saussure 1861: 314. Holotype ♀, Federal District, Mexico. GM. Transferred to *Orphulella* by Bruner 1904: 81.

Orphulella trypha Otte 1979a. Holotype ♀, Martin, British West Indies.
ANSP.

Orphulina balloui (Rehn)

Orphulella balloui Rehn 1905: 178. Holotype ♂, Bay Estate, Barbados,
West Indies. ANSP. Transferred to *Orphulina* by Rehn 1906: 21.

Orphulina acuta Rehn 1906: Holotype ♂, São Paulo, Brazil. ANSP.
Synonym of *Orphulina veteratoria*, Rehn 1918: 194. Synonymized
by Otte 1979a.

Orphulina veteratoria Rehn 1906: 21. Holotype ♀, São Paulo, Brazil. ANSP. Synonymized by Otte 1979a.

Paropomala pallida Bruner 1904: 40. Lectotype ♀, designated by Rehn and Hebard 1912a, Indio, California. ANSP. Transferred to *Eremiacris* by Hebard 1935: 298. Transferred to *Paropomala* by Jago 1969.

Paropomala dissimilis Bruner 1904: 41. Holotype ♀, southern Arizona, California, or northern Mexico. ANSP. Transferred to *Eremiacris* by Hebard 1935.

Paropomala acris Rehn and Hebard 1908: 371. Holotype ♂, Railroad Pass, Cochise Co., Arizona. ANSP. Synonymized by Jago 1969.

Paropomala perpallida Rehn and Hebard 1908: 373. Holotype ♂, near Bright Angel Trail, Grand Canyon. ANSP. Transferred to *Eremiacris* by Hebard 1935: 298. Synonymized by Jago 1969.

Paropomala virgata (Scudder)

Paropomala virgata Scudder 1899c: 437. Lectotype ♂, designated by Jago 1969, Mesilla, New Mexico. ANSP. Transferred to *Eremiacris* by Hebard 1929: 368. Transferred to *Paropomala* by Jago 1969.

Paropomala wyomingensis (Thomas)

Mesops wyomingensis Thomas 1871: 152. Lectotype ♂, designated by Hebard 1927, Cottonwood Creek area, Black Hills, Wyoming. USNM.

Mesops cylindricus Bruner 1890: 48. Lectotype ♂, designated by Rehn and Hebard 1912, Valentine, Nebraska. ANSP. Transferred to *Paropomala* by Scudder 1899. Synonymized by Rehn and Hebard 1906.

Paropomala calamus Scudder 1899: 437. Lectotype ♂, designated by Jago 1969, Lancaster, Calif. ANSP. Synonymized by Rehn and Hebard 1906.

Phaneroturis cupido Bruner 1904: 88. Holotype ♂, Guatemala. GM.

Phaneroturis tantillus Otte 1979b. Holotype ♂, 8.8 road mi W of Los Encuentros on Highway 1, Sololá, Guatemala. UMMZ.

Phlibostroma quadrimaculatum (Thomas)

Stenobothrus quadrimaculatus Thomas 1871a: 266, 280. Holotype ♀, southern Colorado. USNM. Transferred to *Phlibostroma* by Scudder 1875.

Phlibostroma parvum Scudder 1876b: 290. Holotype ♂, northern New Mexico. ANSP. Synonymized by Scudder 1899.

Stenobothrus laetus Uhler 1877: 792. Holotype ♂, Canyon City and Grand Canyon of the Arkansas. ANSP. Synonymized by Jago 1969.

Phlibostroma pictum Scudder 1875a: 517. Holotype ♂, Glencoa, Dodge Co., Nebraska. USNM. Synonymized by McNeill 1897b: 249.

Prorocorypha snowi Rehn 1911: 301. Lectotype ♂ nymph, 5,000–8,000 ft, Santa Rita Mts., Arizona. University of Kansas Museum. Plesiotypes ♂ and ♀, ANSP. Placed under *Mesopsis* by Jago 1969.

Pseudopomala brachyptera (Scudder)

Opomala brachyptera Scudder 1862: 454. Holotype ♂, Princeton, Massachusetts. ANSP. Transferred to *Pseudopomala* by Morse 1896a: 325. Transferred to *Chloealtis* by Jago 1969: 292.

Opomala aptera Scudder 1869a: 305. Holotype ♂, Pennsylvania. According to Jago, 1969, type is lost. Synonymized by Hebard 1925b: 50.

Psoloessa brachyptera (Bruner). NEW COMBINATION

Stirapleura brachyptera Bruner 1905: 105. Holotype ♀, Oaxaca, Mexico. ANSP.

Psoloessa delicatula (Scudder)

Scyllina delicatula Scudder 1876a: 263. Lectotype ♂, designated by Rehn 1942: 207, Garden of the Gods, Colorado. ANSP.

Psoloessa coloradensis Thomas 1876a: 252. ♂ and ♀ types, Denver, Colorado. Types lost, Rehn 1942: 207. Synonymized by McNeill 1897b: 271 under *Stirapleura decussata*.

Dociostaurus ornatus Scudder 1876b: 287. Holotype ♂, northern New Mexico. ANSP. Synonymized by Rehn 1942: 208.

Stirapleura decussata Scudder 1876b: 290. Type ♀, southern Colorado. ANSP. Synonymized by Hebard 1925b: 65.

Psoloessa (?) *eurotiae* Bruner 1890: 62. Lectotype ♂, designated by Rehn and Hebard 1912a, Laramie R. just inside Colorado line. ANSP. Synonymized by Rehn 1942.

Stirapleura tenuicarina Scudder 1899b: 53. Type ♀, Sierra Blanca, Hudspeth Co., Texas. ANSP. Synonymized by Rehn 1942: 208.

Psoloessa delicatula buckelli Rehn 1937: 326. Holotype ♂, Oliver, Okanagan Lake, British Columbia. ANSP.

Psoloessa microptera Otte 1979a. Holotype ♂, Pablillo, Municipio Galeana, Nuevo León, Mexico. ANSP.

Psoloessa salina (Bruner) NEW COMBINATION

Stirapleura salina Bruner 1905: 107. Holotype ♀, Salina Cruz, Mexico. ANSP.

Stirapleura meridionalis Bruner 1905: 107. Holotype ♂, Salina Cruz, Mexico. ANSP. Synonymized by Rehn 1940: 111.

Psoloessa texana Scudder 1875a: 512, Lectotype ♀, designated by Rehn 1942, Dallas, Texas. ANSP.

Psoloessa ferruginea Scudder 1875a: 513. Holotype ♀, Dallas, Texas. ANSP. Synonymized by Rehn and Hebard 1908: 381.

Psoloessa maculipennis Scudder 1875a: 513. Holotype ♀, Dallas, Texas. ANSP. Synonymized by Rehn and Hebard 1908: 381.

Psoloessa buddiana Bruner 1890: 61. Lectotype ♀, designated by Rehn and Hebard 1912, Carrizo Springs, Texas. ANSP. Synonymized by Rehn and Hebard 1908: 381.

Stirapleura pusilla Scudder 1899b: 52. Lectotype ♂, designated by Rehn and Hebard 1912, Mesilla, New Mexico. ANSP. *Psoloessa texana pusilla,* Rehn 1942: 174.

Stirapleura mescalero Rehn 1903: 719. Holotype ♀, Highrolls, New Mexico. ANSP. Synonymized by Rehn and Hebard 1909: 145.

Psoloessa thamnogaea Rehn 1942: 175. Holotype ♀, Murray Canyon above Mission Valley, San Diego Co., California. ANSP. NEW SYNONYM.

Psoloessa texana pawnee Rehn 1942: 206. Holotype ♀, Martin, Bennett Co., South Dakota. ANSP.

Rhammatocerus cyanipes (Fabricius)

Gryllus cyanipes Fabricius 1775: 292. America. Location of type not known.

Stenobothrus gregarius Saussure 1861: 318. Holotype ♂, Saint Thomas, West Indies. GM. Synonymized by Hebard 1924.

Rhammatocerus viatorius (Saussure)

Stenobothrus viatorius Saussure 1861: 317. Lectotype ♂, here designated. The lectotype in the GM was selected by Carbonell on his visit to Geneva in 1970. It bears only the following labels: ''Scyllina viatoria Ss type ♂/Plectrophorus viatorius Sauss *type!*/ Scyllina viatoria Sauss. Hololectotypus ♂ Cs Carbonell 1970.'' Transferred to *Plectrotettix* by Bruner 1904: 101. Transferred to *Scyllina* by Rehn 1906: 48. Transferred to *Rhammatocerus* by Rehn 1940: 103.

Stenobothrus nobilis Walker 1871: 79. Holotype ♂, Oaxaca, Mexico. BM. Synonymized by Hebard 1924.

Plectrotettix patriae Scudder 1901a: 95. Holotype ♀, locality: ''Cala?'' ANSP. Synonymized by Hebard 1924.

Plectrotettix calidus Bruner 1904: 101. Lectotype ♂, designated by Rehn and Hebard 1912a, Cuernavaca, Mexico. ANSP. Synonymized by Hebard 1924.

Plectrotettix macneilli Bruner 1904: 102. Lectotype ♂, designated by Hebard 1924, Orizaba, Veracruz, Mexico. ANSP. Synonymized by Hebard 1924.

Plectrotettix excelsus Bruner 1904: 102. Lectotype ♀, designated by Rehn and Hebard 1912, Tlalpam, Mexico. ANSP. Synonymized by Hebard 1924.

Silvitettix aphelocoryphus Jago 1971: 273. Holotype ♂, San Juanito Mts., 30 mi NE of Tegucigalpa, Tegucigalpa Province, Honduras. ANSP.

Silvitettix audax Otte and Jago 1979. Holotype ♂, Finca La Paz, near La Reforma, San Marcos, Guatemala. UMMZ.

Silvitettix biolleyi (Bruner)

Leuconotus biolleyi Bruner 1904: 57. Holotype ♂, Rio Grande, Costa Rica. ANSP. Transferred to *Silvitettix* by Jago 1971: 280.

Silvitettix chloromerus Jago 1971: 274. Holotype ♂, Fortin, km marker 329, W of Córdoba, Veracruz, Mexico. ANSP.

Silvitettix communis Bruner

Silvitettix communis Bruner 1904: 56. Holotype ♂, Monte Redondo, Costa Rica. ANSP.

Silvitettix hephaistotecnus Jago 1971: 281. Holotype ♂, El Volcan, Santa Rio, Chiriquí, Panama. ANSP. Synonymized by Otte and Jago 1969.

Silvitettix gorgasi (Hebard)

Leuconotus gorgasi Hebard 1924: 97. Holotype ♂, Panama City, Panama. ANSP. Transferred to *Silvitettix* by Jago 1971: 278.

Silvitettix maculatus Otte and Jago 1979. Holotype ♂, 36 road mi S of El Caya, on mountain pine ridge road, Belize. UMMZ.

Silvitettix rhachycoryphus Jago 1971: 275. Holotype ♂, Oaxaca State, Mexico. ANSP.

Silvitettix ricei Otte and Jago 1979. Holotype ♂, 22 mi E of Puerto Juárez, Quintana Roo, Mexico. ANSP.

Silvitettix salinus (Bruner)

Ochrotettix salinus Bruner 1904: 56. Lectotype ♂, designated by Rehn and Hebard 1912, Salina Cruz, Mexico. ANSP. Transferred to *Silvitettix* by Jago 1971: 278.

Silvitettix thalassinus Jago 1971: 279. Holotype ♂, Oricuajo, Rio Jesus Maria, Costa Rica. ANSP.

Silvitettix whitei (Hebard)

Oaxacella whitei Hebard 1932: 232. Holotype ♀, Almoloyas, Oaxaca, Mexico. ANSP. Transferred to *Silvitettix* by Jago 1971: 271.

Stenobothrus (Bruneria) brunneus Thomas

Stenobothrus brunneus Thomas 1871a: 266, 280. Type ♂ and ♀, near Canon City, Colorado on the Arkansas R. (dried alcoholic). Types lost. Transferred to *Platybothrus* by Scudder 1898: 237. Transferred to *Bruneria* by Hebard 1926: 56. See Jago 1971.

Stenobothrus (Bruneria) shastanus (Scudder)

Gomphocerus shastanus Scudder 1880: 25. Lectotype ♂, here designated, Mt. Shasta, California. ANSP. Transferred to *Bruneria* by McNeill 1897. See Jago 1971.

Stenobothrus sordidus McNeill 1897b: 263. Holotype ♂, Salmon City, Idaho, USNM. Transferred to *Platybothrus* by Scudder 1898: 239. Transferred to *Bruneria* by Hebard 1928: 230. NEW SYNONYM.

Platybothrus alticola Rehn 1906: 284. Holotype ♂, Beaver and Piute Cos., Utah, 8,000–10,000 ft. Brooklyn Institute of Arts and Sciences. Transferred to *Bruneria* by Hebard 1926: 56. NEW SYNONYM.

Stenobothrus (Bruneria) yukonensis Vickery

Bruneria yukonensis Vickery 1969b: 265. Holotype ♂, Lake Laberge shore, Yukon, Canada. LEM.

Stethophyma gracile (Scudder)

Arcyptera gracilis Scudder 1862: 462. Lectotype ♂, designated by Otte 1979, Red River or Northern Settlements, Manitoba (now Winnipeg?). ANSP. Transferred to *Stethophyma* by Thomas 1873. Transferred to *Mecostethus* by Morse 1896a: 327.

Arcyptera platyptera Scudder 1863: 463. Holotype ♀, New England. ANSP. Synonymized by Otte 1979b.

Stethophyma lineata (Scudder)

Arcyptera lineata Scudder 1863: 462. Lectotype ♂, here designated, Sanborn, Massachusetts. ANSP. Transferred to *Mecostethus* by Morse 1896. Transferred to *Stethophyma* by Hebard 1926: 59.

Stethophyma celata Otte 1979b. Holotype ♂, Cambridge, Nebraska. ANSP.

Syrbula admirabilis (Uhler)

Stenobothrus admirabilis Uhler 1864: 553. Holotype ♀, Baltimore, Maryland. Type is evidently lost; it is not at ANSP with other species described in same publication. Transferred to *Syrbula* by McNeill 1897b: 221.

Syrbula leucocerca Stål 1873: 102. Holotype ♂, Texas. SM. Synonymized by Jago 1971: 238.

Syrbula festina Otte 1979b. Holotype ♂, 5.5 mi E of Tehuantepec, on Route 190, Oaxaca, Mexico. UMMZ.

Syrbula montezuma (Saussure)

Oxycoryphus montezuma Saussure 1861: 316. Holotype ♂, Cordova, Mexico. GM. Transferred to *Syrbula* by Stål 1873: 102.

Syrbula fuscovittata Thomas 1875: 870. Described from male sex, taken in lower Arizona by Henshaw on Wheeler Expedition of 1874. According to Hebard 1927: 2, type is missing. Synonymized by Otte 1979b.

Syrbula acuticornis Bruner 1890: 55. Lectotype ♀, designated by Rehn and Hebard 1912, southwestern Texas. ANSP. Synonym of *S. fuscovittata,* Hebard 1929: 323.

Syrbula pacifica Bruner 1904: 44. Lectotype ♀, designated by Rehn and Hebard 1912a, Tepic, Mexico. ANSP. NEW SYNONYM.

Syrbula eslavae Rehn 1900: 90. Holotype ♂, Eslava, Federal District, Mexico. ANSP. Subspecies of *S. montezuma,* Hebard 1932: 231.

Syrbula (subgenus *Herus*) *valida* Rehn 1900: 91. Type ♀♀ Eslava, Federal District, Mexico. ANSP. Synonym of *S. eslavae,* Rehn 1906: 18.

Syrbula robusta Bruner 1904: 46. Holotype ♀, Ventanas, Durango, Mexico, 2,000 ft. The type, ''an imperfect specimen,'' is lost, but comes from the geographic region of *S. montezuma.*

Syrbula modesta Bruner 1904: 46. Lectotype ♂, designated by Rehn and Hebard 1912a, Grand Canyon, Colorado R., Arizona. ANSP. Synonym of *S. fuscovittata,* Hebard 1929: 323.

Xeracris minimus (Scudder)

Heliastus minimus Scudder 1900b: 46. Lectotype ♂, here designated, Palm Springs, California. ANSP. Transferred to *Xeracris* by Caudell 1915: 25.

Xeracris snowi (Caudell)

Coniana snowi Caudell 1915: 25. Holotype ♀, Bill Williams Fork, Arizona. USNM. Transferred to *Xeracris* by Jago 1971: 283.

Classification Systems

Classification of certain grasshopper genera according to Rehn and Grant 1960, Dirsh 1975, Uvarov 1966, Jago 1971, and this volume. A name in parentheses indicates the probable classification of a genus based on the diagnostic characters used by the specialist.

Genus	Rehn and Grant	Dirsh	Uvarov	Jago	Otte
Metaleptea	Acridinae	Hyalopteryxinae	(Acridinae)	Acridinae	Acridinae
Orphula	(Acridinae)	Hyalopteryxinae	(Acridinae)	Acridinae	Acridinae
Chrysochraon	Acridinae	Chrysochraotinae	Gomphocerinae	Gomphocerinae	Gomphocerinae
Stethophyma	Acridinae	(Oedipodinae)	Oedipodinae	Acridinae	Gomphocerinae
(*Radinotatum*)[1]	Acridinae	Gomphocerinae	Acridinae	Gomphocerinae	Gomphocerinae
(*Parachloebata*)[2]	Acridinae	?	Acridinae	Acridinae	Gomphocerinae
Cibolacris	Oedipodinae	(Gomphocerinae)	(Gomphocerinae)	Gomphocerinae	Gomphocerinae
(*Coniana*)[3]	Oedipodinae	(Gomphocerinae)	(Gomphocerinae)	Gomphocerinae	Gomphocerinae
Xeracris	Oedipodinae	(Gomphocerinae)	(Gomphocerinae)	Gomphocerinae	Gomphocerinae
Melanotettix	Acridinae	?	(Acridinae)	Gomphocerinae	Gomphocerinae
All other genera	Acridinae	Gomphocerinae	Gomphocerinae	Gomphocerinae	Gomphocerinae

1. Synonym of *Achurum* (Jago 1971).
2. Synonym of *Orphulella* (Otte 1979a).
3. Synonym of *Xeracris* (Jago 1971).

Pronunciation of Names

There is no standard way to pronounce the names of the North American grasshopper genera; pronunciations vary so much that investigators sometimes have difficulty communicating with each other. The pronunciations below seem to be the most common and so are recommended to students.

The vowels have the following sounds: *A* can be sounded either as in *can* (a) or as in *bar* (ah). *E* is either short as in *end* (e) or long as in *peek* (ee). *I* is either short as in *pin*, long as in *peek* (ee), or long as in *pie* (ie). *O* is either short as in *top* (o) or long as in *toe* (oh). *U* is either short as in *push* or long as in *you*. *Y* is always short as in *tin*. Italics indicate which syllables are stressed.

Acantherus Acan*the*rus
Achurum A*koo*rum
Acrolophitus Acrolo*fie*tus
Aeropedellus Airope*del*lus
Ageneotettix Ageneo*tet*tix
Amblytropidia Am*blitrohpi*deeah
Amphitornus Am*phitor*nus
Aulocara Awelo*kah*rah
Boopedon Boh*oh*pedon
Bootettix Boh*oh*tet*tix
Chiapacris Cheeah*pah*criss
Chleoaltis Klee*yal*tiss
Chorthippus Kore*tip*pus
Chrysochraon *Kris*socrayon
Cibolacris Sib*bola*criss
Compsacrella *Komps*ahcrel*lah
Cordillacris Core*dila*criss
Dichromorpha Die*cromore*fah
Eritettix Eri*tet*tix
Esselenia Esse*len*nia
Eupnigodes Yoonig*oh*deez

Heliaula Heeleeawelah
Horesidotes Horesee*doh*teez
Leurohippus Looroh*hip*pus
Melanotettix Melano*tet*tix
Mermiria Mur*mireeah
Opeia Oh*pie*yah
Orphula Or*few*lah
Orphulella Orfew*lel*lah
Orphulina Orfew*lee*nah
Paropomala Pah*rohpoh*mahlah
Phaneroturis Fanerohtooriss
Phlibostroma Fli*bosstroh*mah
Prororocorypha Prohrohcorifah
Pseudopomala Soo*dohpoh*mahlah
Psoloessa Sol*es*sa
Rhammatocerus Ram*mato*cerus
Silvitettix Sil*vitet*tix
Stenobothrus Stenoh*both*rus
Stethophyma Stetho*fie*mah
Syrbula Sir*bewlah
Xeracris Zeer*ac*ris

Glossary

Accessory carinae (or *carinulae*): two distinct or indistinct ridges between the median carina and the lateral carinae; when the latter are missing it may be difficult to distinguish between accessory and lateral carinae on the pronotum; usually the former are located near the median carina (Figs. 26, 29, 33).

Allopatry: two or more populations or species living in separated geographic areas (in contrast to parapatry and sympatry).

Anterior: located toward the front.

Apterous: without wings.

Arcuate groove of fastigium: the curved groove that runs across the fastigium (Fig. 18A, D).

Brachypterous: with short, flightless wings.

Carina: an anatomical ridge.

Carinula: a small or low anatomical ridge, sometimes a barely raised line.

Carinulae of frontal ridge: raised lateral margins of the frontal ridge (Fig. 6).

Cerci: two small conical or flattened appendages at the end of the abdomen (Fig. 5).

Crepitation: communicative snapping sounds made with the wings during flight.

Disk of pronotum: the flat area on top of the pronotum (Fig. 7).

Dorsal: on the upper surface of the body.

Dorsal field of forewing: the horizontal area of the forewing (Fig. 5).

Dorsum: upper surface of the body.

Ensiform: sword-shaped; usually referring to antennae that are flattened at the base of the flagellum, then taper gradually (in contrast to filiform) (Fig. 19A).

Epiproct: the dorsal plate between the bases of the cerci (Fig. 5).

Fastigial carinae: ridges bordering the front and sides of the fastigium.

Fastigium: depression or flattened area at the extreme top, front, and center of the head (Fig. 6).

FDE or *fascia dorsalis exterior:* darkened regions on the upper sides of the lateral lobes and adjacent to the lateral carinae (Fig. 45D).

FDI or *fascia dorsalis interior:* dark marking (often triangular in shape) on the pronotal disk adjacent to the lateral carinae (Fig. 40A, B, C).

Flagellum: the main portion of the antennae, excluding the basal segment or scape (Fig. 5).

Foveolar area: the area of the head where the lateral foveolae normally are located (Fig. 6).

Frons: dorso-anterior extremity of the head.

Frontal ridge or frontal costa: raised area along the midline of the face; sometimes grooved, forming two carinulae (Fig. 6).

Holotype: the single specimen designated or indicated as the type in the original published description.

Homonym: single name given to two or more species. The first species so named retains the name; subsequent species so named are renamed.

Junior synonym: see *Synonym.*

Knee: the thickened end of the hind femur (Fig. 8).

Lateral carinae: the paired ridges usually running along and defining the lateral edges of the pronotal disk (Fig. 7).

Lateral field of forewing: the vertical surface of the forewing (Fig. 5).

Lateral foveolae: small depressions on either side of the fastigium at the top and front of the head (Fig. 6).

Lateral lobes of pronotum: the vertical side extensions of the pronotum (Fig. 7).

Lectotype: the one specimen from a type series subsequently chosen to be the type.

Median carina: the median ridge, often running the entire length of the pronotum, but sometimes distinct only in parts; sometimes an indistinct raised line (Fig. 7).

Mesosternum: the ventral side of the second thoracic segment.

Metazona: the region of the pronotum in back of the principal sulcus (Fig. 7).

Neotype: a specimen selected as the type subsequent to the original description if the original type is lost or has been destroyed, or suppressed by the International Commission on Zoological Nomenclature.

Occiput: the top and back of the head where it meets the pronotum.

Parapatry: the occurrence of two or more species in adjacent but nonoverlapping geographical areas.

Paratype: a specimen other than the holotype that the author had at the time of the original description and so designated.

Posterior: located toward the rear.

Postocular stripe: darkened band running from the back of the eye posteriorly across the top of the pronotum and often extending onto the thorax and forewing (Figs. 22E, F, G; 30A, B).

Preocular ridge: the distinct or indistinct vertical ridge on the face between the antennal socket and the compound eye.

Principal sulcus: the rear transverse sulcus on the pronotum, usually the deepest and best-defined sulcus (Fig. 7).

Pronotum: the first large plate behind the head, consisting of a horizontal region (the disk) and two vertical side flaps (the lateral lobes).

Prosternal spine: a spine or projection extending down between the front legs (Figs. 15, 22A).

Prosternum: the ventral side of the first thoracic segment.

Prozona: the region of the pronotum anterior to the principal (or last) transverse sulcus (Fig. 7).

Scape: the basal segment of the antenna.

Sclerite: a single, more or less rigid exoskeletal plate.

Senior synonym: see *Synonym.*

Spines: immovable projections on the tibiae.

Spurs: movable spinelike projections on the legs.

Sternite: a single segmental plate on the venter of the body.

Sternum: the ventral segments of the body.

Stridulation: the sound produced by the rubbing of one part of the exoskeleton against another part.

Stridulatory pegs or teeth: the small teeth located on the inner lower carinula of the hind femur (Fig. 21A).

Subgenital plate: the last sternite of the male abdomen.

Subocular groove: the vertical groove or depression running from the bottom of the eye to the anterior base of the mandible (Fig. 6).

Sulcus (plural *sulci*): groove(s) or narrow depression(s) across the pronotum or vertically on the lateral lobes.

Suture: the juncture between two exoskeletal plates (or sclerites).

Sympatry: the occurrence of two or more species in the same area (in contrast to allopatry and parapatry).

Synonym: each of two or more different names for the same taxon; the *senior synonym* is the earliest published name; a *junior synonym* is a more recently published name.

Tergite: a single plate on the dorsum of the body.

Tergum: the dorsal segments of the body.

Tympanum: the hearing organ at the base of the abdomen, usually above the base of the hind leg and often hidden beneath the forewing (Fig. 5).

Venter: the lower part or underside of the body or a body part.

Ventral: on the lower side of the body.

Vertex: the highest part of the head (Fig. 6).

References

Alexander, G., and J. R. Hilliard. 1964. Life history of *Aeropedellus clavatus* (Orthoptera: Acrididae) in the Alpine tundra of Colorado. *Annals of the Entomological Society of America* 57: 310–317.

Alexander, G., and H. G. Rodeck. 1952. Two species of Great Basin Orthoptera new to Colorado. *Entomological News* 63: 238–240.

Alexander, R. D., and D. Otte. 1967. The evolution of genitalia and mating behavior in crickets (Gryllidae) and other Orthoptera. *Miscellaneous Publications of the University of Michigan Museum of Zoology* 133: 1–62.

Amedegnato, C. 1974. Les genres d'acridiens Neotropicaux, leur classification par familles, sous-familles et tribus. *Acrida* 3: 193–204.

——— 1977. Etude es *Acridoidea* Centre et Sud Americains (Catantopinae sensulato). Anatomie des genitalia, classification, repartition, phylogenie. Doctoral dissertation, University of Pierre and Marie Curie, Paris.

Anderson, N. L., and J. C. Wright. 1952. Grasshopper investigations on Montana range lands. *Montana State College Agricultural Experimental Station Bulletin* 486: 46.

Ball, E. D., E. R. Tinkham, R. Flock, and C. T. Vorhies. 1942. The grasshoppers and other Orthoptera of Arizona. *Technical Bulletin Arizona College of Agriculture* 93: 275–373.

Bei-Bienko, G. Y. 1932. The group Chrysochraontes (Acrididae). *Eos* (Madrid) 8: 43–92.

Bei-Bienko, G. Y., and L. L. Mishchenko. 1964. Locusts and grasshoppers of the U. S. S. R. and adjacent countries. Part II. *Zoological Institute of U. S. S. R. Academy of Sciences* no. 40.

Blatchley, W. S. 1902. *A Nature wooing at Ormond by the Sea.* Indianapolis.

——— 1920. *Orthoptera of northeastern America.* Indianapolis.

Bolivar, I. 1888. Enumération des Orthopteres de l'île de Cuba. *Memoires Société Zoologie France* 1: 116–164.

Brooks, A. R. 1958. Acridoidea of southern Alberta, Saskatchewan and Manitoba (Orthoptera). *Canadian Entomologist* 90, suppl. 9.

Bruner, L. 1885a. Contributions of the North Trans-Continental Survey. Orthoptera. *Canadian Entomologist* 17: 9–19.

———— 1885b. First contribution to a knowledge of the Orthoptera of Kansas. *Bulletin Washburn College Laboratory National History* 1: 125–139.

———— 1890 (1889 vol.). New North American Acrididae, found north of the Mexican boundary. *Proceedings U. S. National Museum* 12: 47–82.

———— 1893. A list of Nebraska Orthoptera. *Publications of the Nebraska Academy of Sciences* 3: 19–33.

———— 1895. Nicaraguan Orthoptera. *Bulletin from the Laboratories of Natural History of the State University of Iowa* 3: 59–69.

———— 1897. The grasshoppers that occur in Nebraska. *Annual Report Entomology, Nebraska State Building of Agriculture*, 1896: 105–138.

———— 1900. A brief account of the genera and species of locusts or grasshoppers of Argentina, together with descriptions of new forms. *Second Report, Merchants Locust Investigation Commission, Buenos Aires*, pp. 1–80.

———— 1900–1909. The Acrididae. *Biologia Centrali-Americana* 2: 19–342.

———— 1906a. Report on the Orthoptera of Trinidad, West Indies. *Journal of the New York Entomological Society* 14: 135–165.

———— 1906b. Some Guatemalan Orthoptera with descriptions of five new species. *The Ohio Naturalist* 7: 9–13.

———— 1911. South American Acridoidea. *Annals of the Carnegie Museum* 8: 5–147.

Brunner, C. de Wattenwyl. 1893. Révision du système des Orthoptères et description des espèces rapportées par M. Leonardo Fea de Birmanie. *Annali del Museo civico di Storia Naturale di Genova* (2) 13: 121.

Buckell, E. R. 1922. A list of the Orthoptera and Dermaptera recorded from British Columbia prior to the year 1922, with annotations. *Proceedings Entomological Society of British Columbia* 20: 9–41.

Burmeister, H. 1838. *Handbuch der Entomologie, Orthoptera*, vol. 2, part 2. Berlin, pp. 397–756.

Campbell, J. B., W. H. Arnett, J. D. Lambley, O. K. Jantz, and H. Knutson. 1974. Grasshoppers (Acrididae) of the Flint Hills native tallgrass prairie in Kansas. *Kansas State University Agricultural Experiment Station Research Paper* no. 19: 1–146.

Cantrall, I. J. 1943. The ecology of the Orthoptera and Dermaptera of the George Reserve, Michigan. *Miscellaneous Publications of the University of Michigan Museum of Zoology* 54: 1–184.

Caudell, A. N. 1903a. Some new or unrecorded Orthoptera from Arizona. *Proceedings of the Entomological Society of Washington* 5(2): 162–166.

———— 1903b. Notes on Orthoptera from Colorado, New Mexico, Arizona, and Texas with descriptions of new species. *Proceedings of the United States National Museum* 26: 775–809.

———— 1904. Orthoptera from southwestern Texas. *Science Bulletin of the Brooklyn Institute of Arts and Sciences* 1: 105–116.

———— 1909. Miscellaneous notes on Orthoptera. *Proceedings of the Entomological Society of Washington* 11: 111–114.

———— 1915. Notes on some United States grasshoppers of the family Acrididae. *Proceedings of the United States National Museum* 49: 25–31.

———— 1922. Report on the Orthoptera and Dermaptera collected by the Barbados-Antigua expedition from the University of Iowa in 1918. *University of Iowa Studies in Natural History* 10 (1): 19–44.

Cohn, T. J., and I. J. Cantrall. 1974. Variation and speciation in the grasshoppers of the Conalcaeini (Orthoptera: Acrididae: Melanoplinae): The lowland forms of western Mexico, the genus Barytettix. *San Diego Society of Natural History Memoir* 6: 1–131.

Coppock, S. 1962. The grasshoppers of Oklahoma (Orthoptera: Acrididae). Ph.D. thesis, Oklahoma State University.

Criddle, N. 1933 (published 1935). Studies in the biology of North American Acrididae, development and habits. *Proceedings, World's Grant Exhibition and Conference, Canada* 2: 474–494.

Dakin, M. E. 1979. Notes on *Chrysochraon* (*Barracris*) *petraea* (Gurney, Strohecker, and Helfer) (Orthoptera, Acrididae, Gomphocerinae) with a description of the female sex. *Acrida* 8: 9–15.

DeGeer, C. 1773. Memoires pour servir a l'histoire des insectes. Stockholm.

Dirsch, V. M. 1961. A preliminary revision of the families and subfamilies of Acridoidea (Orthoptera, Insecta). *Bulletin of the British Museum (Natural History), Entomology.* 10: 351–419.

———— 1975. *Classification of the Acridomorphoid Insects*. Oxford.

Faber, A. 1953. Laut- und Gebärdensprache bei Insekten. Orthoptera (Geradflügler). I. Mitteilungen der Staatliche Museum für Naturkunde, Stuttgart: 1–198.

Fabricius, Johann C., 1775. *Systema entomologiae, sistens insectorum, classes, ordines, genera, species, adiectis, synonymis, locis, descriptionibus, observationibus*. Flensburgi et Lipsiae, in officina libraria Kortii, 1775.

Fieber, C. 1852. Grundlage zur Kenntniss der Orthopteren (Geradflügler) Oberschlesiens und Grundlage zur Kenntniss der Kafer Oberschlesiens. *Naturwissenschaftliche Zeitschrift* (3) 1852: 100.

Fisher, L. H. 1853. Orthoptera Europaea: 1–454.

Fox, H. 1914. Data on the Orthopteran faunistics of eastern Pennsylvania and southern New Jersey. *Proceedings of The Academy of Natural Sciences of Philadelphia* 1914: 441–534.

Giglio-Tos, E. 1894. Viaggio del dott. Alfredo Borelli nella Republica Argentina e nel Paraguay. VII. Ortotteri. *Bolletino del Musei di Zoologia ed Anatomia Comparata della R. Università di Torino* 9 (184): 1–46.

———— 1897. Ortotteri—raccolti nel Darien cal Dr. E. Festa. III. Acrididae —Gryllidae. *Bolletino del Musei di Zoologia ed Anatomia Comparata dell R. Università di Torino* 12 (301): 1–10.

Gurney, A. B. 1940. A revision of the grasshoppers of the genus *Orphulella* Giglio-Tos, from America north of Mexico (Orthoptera: Acrididae). *Entomologica Americana* 20 (3): 85–157.

———— 1959. A new grasshopper of the genus *Achurum* from eastern Texas (Orthoptera: Acrididae). *Journal of the Washington Academy of Sciences* 49: 117–120.

Gurney, A. B., H. F. Strohecker, and J. R. Helfer. 1964. A synopsis of the North American acridine grasshoppers of the genus group *Chrysochraontes* (Orthoptera: Acrididae). *Transactions of the American Entomological Society* 89: 119–137.

Hancock, J. L. 1906. On the Orthopteran genus *Ageneotettix* with a description of a new species from Illinois. *Entomological News* 17: 253–256.

Harris, T. W. 1935. A catalogue of the animals and plants in Massachusetts. VIII. Insects. *In* E. Hitchcock, *Report on the geology, minerology, botany and zoology of Massachusetts*. 2nd ed. Amherst, Mass., pp. 553–602.

———— 1841. *A report on the insects of Massachusetts, injurious to vegetation*. Cambridge. Mass.

Hart, C. A. 1906. Notes of a winter trip in Texas, with an annotated list of the Orthoptera. *Entomological News* 17: 154–160.

Haskell, P. T. 1956. Hearing in certain Orthoptera. II. The nature of the response of certain receptors to natural and imitation stridulation. *Journal of Experimental Biology* 33: 767–776.

———— 1961. *Insect Sounds*. Chicago.

Hebard, M. 1920. A new genus and species of grasshopper from California. *Proceedings of the California Academy of Sciences* 10: 71–75.

———— 1922. North American Acrididae (Orthoptera). Paper 1. *Transactions of the American Entomological Society* 48: 89–109.

———— 1923. Studies in the Dermaptera and Orthoptera of Colombia. *Transactions of the American Entomological Society* 49: 165–313.

———— 1924a. Studies in the Acrididae of Panama (Orthoptera). *Transactions of the American Entomological Society* 50 (851): 75–140.

———— 1924b. Studies in the Dermaptera and Orthoptera of Ecuador. *Proceedings of The Academy of Natural Sciences of Philadelphia* 76: 109–248.

———— 1925a. Dermaptera and Orthoptera from the state of Sinaloa, Mexico. Part II. *Transactions of the American Entomological Society* 51: 265–310.

———— 1925b. The Orthoptera of South Dakota. *Proceedings of the Academy of Natural Sciences of Philadelphia* 77: 33–155.

———— 1926. A key to the North American genera of the Acridinae which occur north of Mexico. Orthoptera (Acrididae). *Transactions of the American Entomological Society* 52 (886): 47–59.

———— 1927. Fixation of the single types of species of Orthoptera described by Cyrus Thomas. *Proceedings of the Academy of Natural Sciences of Philadelphia* 79: 1–11.

———— 1928. The Orthoptera of Montana. *Proceedings of the Academy of Natural Sciences of Philadelphia* 80:211–306.

———— 1929. The Orthoptera of Colorado. *Proceedings of the Academy of Natural Sciences of Philadelphia* 81: 303–425.

—— 1931. Studies in lower Californian Orthoptera. *Transactions of the American Entomological Society* 57: 113–127.

—— 1932. New species and records of Mexican Orthoptera. *Transactions of the American Entomological Society* 58: 201–371.

—— 1933. Studies in the Dermaptera and Orthoptera of Colombia. *Transactions of the American Entomological Society* 49: 13–67.

—— 1935. Notes on the Group Gomphoceri and a key to its genera, including one new genus (Orthoptera, Acrididae, Acridinae). *Entomological News* 46: 184, 204–208.

—— 1937. Studies in Orthoptera which occur in North America north of the Mexican boundary. *Transactions of the American Entomological Society* 63: 347–379.

Hubbell, T. H. 1922. Notes on the Orthoptera of North Dakota. *Occasional Papers, Museum of Zoology, University of Michigan* 113: 1–56.

—— 1954. The naming of geographically variant populations. *Systematic Zoology* 3: 113–121.

Isely, F. B. 1937. Seasonal succession, soil relations, numbers, and regional distribution of northeastern Texas acridians. *Ecological Monographs* 7: 317–344.

Jacobs, W. 1953. Verhaltensbiologische Studien an Feldheuschrecken. *Zeitschrift für Tierpsychologie*. Supplement 1: 1–228.

Jago, N. D. 1969. A revision of the systematics and taxonomy of certain North American Gomphocerine grasshoppers (Gomphocerinae, Acrididae, Orthoptera). *Proceedings of the Academy of Natural Sciences of Philadelphia* 121 (7): 229–335.

—— 1971. A review of the Gomphocerinae of the world with a key to the genera (Orthoptera, Acrididae). *Proceedings of the Academy of Natural Sciences of Philadelphia* 123: 205–343.

Joern, A. 1979. Resource utilization and community structure in assemblages of arid grassland grasshoppers (Orthoptera: Acrididae). *Transactions of the American Entomological Society* 105: 253–300.

Johannson, B., 1763. Centurio insectorum quam praeside D. D. Car. von Linne propofuit B. J. Calmarienfis. In Carolia von Linne, *Amoenitates Academicae seu dissertationes variae Physicae, Medicae, Botanicae anthehac seorsim editae*, vol 6. 2nd ed. 1789, Erlanger.

Kevan, D. K. McE. 1954. Méthodes inhabituelles de production de son chez les orthoptères. In Busnel, R. G. (ed.), *L'Acoustique des Orthoptères*. Paris.

Kirby, W. F. 1910. *A synonymic catalogue of Orthoptera*, vol. 3. British Museun of Natural History Publication.

Kreasky, J. B. 1960. Extended diapause in eggs of high-altitude species of grasshoppers, and a note on food-plant preferences of *Melanoplus bruneri*. *Annals of the Entomological Society of America* 53: 436–438.

McNeill, J. 1897a. Some corrections in generic names in Orthoptera. *Psyche* 8: 71.

—— 1897b. Revision of the Truxalinae of North America. *Proceedings of the Davenport Academy of Natural Sciences* 6: 179–274.

————— 1899. Notes on Arkansas Truxalinae. *Canadian Entomologist* 31: 53–55.

Mayr, E. 1963. *Animal Species and Evolution.* Cambridge, Mass.

Michelsen, A. 1966. Pitch discrimination in the locust ear: observations on single sense cells. *Journal of Insect Physiology* 12: 1119–1131.

Morse, A. P. 1893. A new species of *Stenobothrus* from Connecticut with remarks on other New England species. *Psyche* 6: 477–479.

————— 1896a. Notes on the Acrididae of New England. II. Truxalinae. *Psyche* 7: 323–327, 342–344, 382–384, 402–403, 407–411, 413–422, 443–445.

————— 1896b. Some notes on locust stridulation. *Journal New York Entomological Society* 4: 16–20.

————— 1903. New Orthoptera from Nevada. *Psyche* 10: 115–116.

————— 1904. Researches on North American Acrididae. *Publication of the Carnegie Institute of Washington* 18: 1–55.

————— 1907. Further researches on North American Acrididae. *Carnegie Institution of Washington* 68: 1–54.

————— 1920. Manual of the Orthoptera of New England. *Proceedings of the Boston Society of Natural History* 35: 197–556.

Morse, A. P., and M. Hebard. 1915. Fixation of single type (Lectotypic) specimens of species of American Orthoptera (described by Morse). *Proceedings of the Academy of Natural Sciences of Philadelphia* 1915: 96–106.

Mulkern G. B., K. P. Pruess, H. Knutson, H. F. Hagen, J. B. Campbell, and J. D. Lambley. 1969. Food habits and preferences of grassland grasshoppers of the North Central Great Plains. *Agricultural Experimental Station North Dakota State University Bulletin* no. 481: 1–31.

Otte, D. 1970. A comparative study of communicative behavior in grasshoppers. *Miscellaneous Publications of the University of Michigan Museum of Zoology* no. 141: 1–168.

————— 1972. Simple versus elaborate behavior in grasshoppers. An analysis of communication in the genus *Syrbula*. *Behaviour* 42: 291–322.

————— 1979a. Revision of the grasshopper tribe Orphulellini (Acrididae, Gomphocerinae). *Proceedings of the Academy of Natural Sciences of Philadelphia* 131: 52–88.

————— 1979b. Descriptions of new North American Gomphocerine grasshoppers (Gomphocerinae: Acrididae). *Proceedings of the Academy of Natural Sciences of Philadelphia* 131: 231–243.

Otte, D., and N. D. Jago, 1979. Revision of the grasshopper genera *Silvitettix* and *Compsacris* (Gomphocerinae). *Proceedings of the Academy of Natural Sciences of Philadelphia* 131: 257–288.

Otte, D., and J. A. Joern. 1975. Insect territoriality and its evolution: population studies of desert grasshoppers on creosote bushes. *Journal of Animal Ecology* 44: 29–54.

————— 1977. On feeding patterns in desert grasshoppers and the evolution of specialized diets. *Proceedings of the Academy of Natural Sciences of Philadelphia* 128: 89–126.

Otte, D., and K. Williams. 1972. Environmentally induced color dimorphisms in grasshoppers, *Syrbula admirabilis, Dichromorpha viridis* and *Chortophaga viridifasciata. Annals of the Entomological Society of America* 65: 1154–1161.

Palisot de Beauvoir, A. M. F. J. 1807. *Insectes recueillis en Afrique et en Amerique dans les royaumes d'Oware et de Benn è Saint-Domingue et dans les Etats-Unis pendant les années 1786–1979.* Paris.

Rehn, J. A. G. 1900. Notes on Mexican Orthoptera, with descriptions of new species. *Transactions of the American Entomological Society* 27: 85–99.

———— 1901. Some necessary changes and corrections in names of Orthoptera. *The Canadian Entomologist* 33: 271–272.

———— 1902a. A contribution to the knowledge of the Orthoptera of Mexico and Central America. *Transactions of the American Entomological Society* 29: 1–34.

———— 1902b. Notes on some generic names employed by Serville, in the Revue Methodique, and Fieber, in the Synopsis die Europaischen Orthopteran. *Canadian Entomologist* 34: 316–317.

———— 1902c. Notes on the Orthoptera of New Mexico and western Texas. *Proceedings of the Academy of Natural Sciences of Philadelphia* 1902: 719.

———— 1904a. Notes on Orthoptera from Northern and Central Mexico. *Proceedings of the Academy of Natural Sciences of Philadelphia* 1904: 513–549.

———— 1904b. Notes on the Orthoptera of the Keweenaw Bay Region of Baraga County, Michigan. *Entomological News* 15: 229–236, 263–270.

———— 1905. Notes on a small collection of Orthoptera from the Lesser Antilles, with the description of a new species of *Orphulella. Entomological News* 16: 173–182.

———— 1906a. Notes on South American grasshoppers of the subfamily (Acridinae, Acrididae), with descriptions of new genera and species. *Proceedings of the U. S. National Museum* 30: 371–391.

———— 1906b. Studies in South and Central American Acridinae (Orthoptera) with the descriptions of a new genus and six new species. *Proceedings of the Academy of Natural Sciences of Philadelphia* 1906: 10–50.

———— 1906c. Some Utah Orthoptera. *Entomological News* 17: 284–288.

———— 1907. Notes on Orthoptera from southern Arizona, with descriptions of new species. *Proceedings of the Academy of Natural Sciences of Philadelphia* 59: 24–81.

———— 1911. Orthoptera from the Santa Rita Mountains, Arizona. Collected by the University of Kansas Expedition. *The Kansas University Science Bulletin* 5 (17): 299–307.

———— 1916. The Stanford Expedition to Brazil, 1911. Dermaptera and Orthoptera I. *Transactions of the American Entomological Society* 42: 215–308.

———— 1917. On Orthoptera from the vicinity of Rio de Janeiro. *Transactions of the American Entomological Society* 43: 335–363.

———— 1918. On Dermaptera and Orthoptera from southeastern Brazil. *Transactions of the American Entomological Society* 44: 181–222.

———— 1919. A study of the orthopterous genus *Mermiria* Stål. *Proceedings of the Academy of Natural Sciences of Philadelphia* 71: 55–120.

———— 1923. A study of the Ligurotettigi. *Transactions of the American Entomological Society* 49: 43–92.

———— 1927. On new and certain previously known American genera of the Acridinae with specific comments and descriptions (Orthoptera, Acrididae). *Transactions of the American Entomological Society* 53: 213–240.

———— 1928. On the relationship of certain new or previously known genera of the Acridinae group Chrysochraontes (Orthoptera, Acrididae). *Proceedings of the Academy of Natural Sciences of Philadelphia* 80: 189–205.

———— 1937. A new subspecies of *Psoloessa delicatula* (Orthoptera: Acrididae). *Transactions of the American Entomological Society* 63: 325–332.

———— 1940. The application, relationship and species of *Scyllina* Stål, 1861, and *Scyllinops* Rehn, 1927 (Orthoptera: Acrididae, Acridinae). *Transactions of the American Entomological Society* 66: 101–120.

———— 1942. On the locust genus *Psoloessa* (Orthoptera, Acrididae, Acridinae). *Transactions of the American Entomological Society* 68: 167–237.

———— 1944. A revision of the locusts of the group Hyalopteryges. *Transactions of the American Entomological Society* 70: 181–234.

Rehn, J. A. G., and H. J. Grant, Jr. 1959. Critical remarks on a recent contribution to the taxonomy of the Acridoidea (Orthoptera) by V. M. Dirsh. *Entomological News* 70: 245–249.

———— 1960. A new concept involving the subfamily Acridinae (Orthoptera, Acridoidea). *Transactions of the American Entomological Society* 86: 173–185.

———— 1960. *Prororcorypha, Eremiacris, Paropomala*. Members of the subfamily Acridinae (*Sensu* Rehn and Grant) (Orthoptera, Acrididae). *Notulae Naturae* no. 336: 1–2.

Rehn, J. A. G., and M. Hebard. 1905. A contribution to the knowledge of the Orthoptera of south and central Florida. *Proceedings of the Academy of Natural Sciences of Philadelphia* 57: 29–55.

———— 1906. A contribution to the knowledge of the Orthoptera of Montana, Yellowstone Park, Utah and Colorado. *Proceedings of the Academy of Natural Sciences of Philadelphia* 58: 358–418.

———— 1907. Orthoptera from northern Florida. *Proceedings of the Academy of Natural Sciences of Philadelphia* 59: 279–319.

———— 1908. An orthopterological reconnaissance of the southwestern United States. Part I: Arizona. *Proceedings of the Academy of Natural Sciences of Philadelphia* 60: 111–175.

———— 1909. An orthopterological reconnaissance of the southwestern United States. Part II: New Mexico and western Texas. *Proceedings of the Academy of Natural Sciences of Philadelphia* 61: 111–175.

———— 1910. An Orthopterological reconnaissance of the southwestern United States. Part III: California and Nevada. *Proceedings of the Academy of Natural Sciences of Philadelphia* 61: 409–483.

———— 1912a. Fixation of single type (lectotype) specimens of species of American Orthoptera. Section one. *Proceedings of the Academy of Natural Sciences of Philadelphia* 64: 60–128.

———— 1912b. On the Orthoptera found on the Florida Keys and in extreme southern Florida. I. *Proceedings of the Academy of Natural Sciences of Philadelphia* 64: 235–276.

———— 1916. Studies in the Dermaptera and Orthoptera of the coastal plain and piedmont region of the southeastern United States. *Proceedings of the Academy of Natural Sciences of Philadelphia* 68: 87–314.

———— 1919. A new species of grasshopper of the genus *Chloealtis* (Acridinae) from the Pacific Slope. *Transactions of the American Entomological Society* 45: 81–87.

———— 1938. New genera and species of West Indian Acrididae with notes on previously-known species (Orthoptera). *Transactions of the American Entomological Society* 64: 201–226.

Rentz, D. C. 1966. A review of the genus *Esselenia* Hebard with the description of a new subspecies (Orthoptera: Acrididae). *Occasional Papers of the California Academy of Sciences* no. 54: 1–10.

Roberts, H. R. 1937. Studies on the family Acrididae (Orthoptera) of Venezuela. *Proceedings of the Academy of Natural Sciences of Philadelphia* 89: 343–368.

———— 1941a. A comparative study of the subfamilies of the Acrididae (Orthoptera) primarily on the basis of their phallic structures. *Proceedings of the Academy of Natural Sciences of Philadelphia* 93: 201–246.

———— 1941b. Nomenclature in the Orthoptera concerning genotype designations. *Transactions of the American Entomological Society* 67: 1–34.

Saussure, H. de. 1861. Orthoptera Nova Americana (Diagnoses Praeliminares) (series IIa). *Revue et Magazin de Zoologie, pure et appliqué* 13: 313–329.

Say, T. 1825a. Descriptions of new Hemipterous insects collected in the expedition to the Rocky Mountains, performed by order of Mr. Calhoun, secretary of war, under command of Major Long. *Journal of the Academy of Natural Sciences of Philadelphia* 4: 307–345.

———— 1825b. American entomology or descriptions of the insects of North America. *Philadelphia Museum Publications*, vol. 2.

Scudder, S. H. 1862. Materials for a monograph of the North American Orthoptera. Including a catalogue of the known New England species. *Journal of Boston Society of Natural History* 7: 409–480.

———— 1869a. Descriptions of new species of Orthoptera in the collection of the American Entomological Society. *Transactions of the American Entomological Society* 2: 305–307.

————— 1869b. Notes on Orthoptera collected by Prof. Jas. Orton on either side of the Andes of Equatorial South America. *Proceedings of the Boston Society of Natural History* 12: 330–345.

————— 1872. Notes on the Orthoptera collected by Dr. F. V. Hayden in Nebraska, In F. V. Hayden, *Final Report of the U. S. Geological Survey of Nebraska and portions on adjacent territories made under the direction of the commissioner of the General Land Office,* pp. 249–261.

————— 1874. The distribution of insects in New Hampshire. *Hitchcock's Report on the Geology of New Hampshire,* Concord, N. H., pp. 331–380.

————— 1875a. A century of Orthoptera. Decade IV—Acrydii. *Proceedings of the Boston Society of Natural History* 17: 510–517.

————— 1875b. Notice on the butterflies and Orthoptera collected by Mr. George M. Dawson, as naturalist of the B. N. A. Boundary Commission. Appendix D in George M. Dawson, *British North American Boundary Commission Geological Report of Progress for the year 1873 (in part). Report of the Tertiary Lignite Formation in the vicinity of the 49th Parallel.* Montreal.

————— 1876a. List of the Orthoptera collected by Dr. A. S. Packard in Colorado and the neighboring territories, during the summer of 1875. *Bulletin of the Geological and Geographical Survey of the Territories* 2 (3): 261–267.

————— 1876b. Report on the Orthoptera collected by the U. S. Geological surveys west of the 100th Meridian under the direction of Lieutenant George M. Wheeler, during the season of 1875. Appendix H9 in G. M. Wheeler, *Annual report upon the Geological surveys west of the 100th Meridian, in California, Nevada, Utah, Colorado, Wyoming, New Mexico, Arizona, and Montana.* Appendix JJ of the *Annual Report of the Chief of Engineers for 1876.*

————— 1877. New forms of saltatorial Orthoptera from the southern United States. *Proceedings of the Boston Society of Natural History* 19: 35–41.

————— 1878a (vol. for 1876–1878) A century of Orthoptera. Decade VII—Acrydii. *Proceedings of the Boston Society of Natural History* 19: 27–35.

————— 1878b. The Florida Orthoptera collected by Mr. J. H. Comstock. *Proceedings of the Boston Society of Natural History* 19: 80–94.

————— 1880. List of Orthoptera collected by Dr. A. S. Packard, Jr., in the western United States in the summer of 1877. Appendix II, *Second Report of the United States Entomological Commission,* pp. 23–28.

————— 1890. Some genera of Oedipodidae rescued from the Tryxalidae. *Psyche* 5: 431–442.

————— 1897. Biological and other notes on American Acrididae. *Psyche* 8: 99–102.

————— 1898. A preliminary classification of the Tryxalinae of the United States and Canada. *Psyche* 7: 231–239.

————— 1899a. The North American species of *Orphulella. Canadian Entomologist* 31: 177–188.

—— 1899b. Short studies in North American Tryxalinae. *Proceedings of the American Academy of Arts and Sciences* 35: 39–57.

—— 1899c. *Pseudopomala* and its allies. *Psyche* 8: 436–438.

—— 1900a. A list of the Orthoptera of New England. *Psyche* 9: 99–106.

—— 1900b. The species of the Oedipodinae genus *Heliastus* Sauss., occurring in the United States. *Psyche* 9: 45–47.

—— 1901a. Alphabetical index to North American Orthoptera described in the eighteenth and nineteenth centuries. *Occasional Papers of the Boston Society of Natural History* 6: 1–436.

—— 1901b. Catalogue of the described Orthoptera of the United States and Canada. *Proceedings of the Davenport Academy of Natural Sciences* 8: 1–101.

Scudder, S. H., and T. D. A. Cockerell. 1902. A first list of the Orthoptera of New Mexico. *Proceedings of the Davenport Academy of Natural Sciences* 9: 1–60.

Serville, J. G. 1839. *Histoire Naturelle des Insects. Orthopteres*. Paris.

Somes, M. P. 1914. The Acridiidae of Minnesota. *Minnesota Agricultural Experiment Station Bulletin* 141: 1–98.

Stål, C. 1873. *Recensio Orthopterorum. Revue critique des Orthopteres decrits par Linne, DeGeer et Thunburg*. Part 1.

Strohecker, H. F., W. W. Middlekauff, and D. C. Rentz. 1968. The grasshoppers of California (Orthoptera, Acridoidea). *Bulletin of the California Insect Survey* 10: 1–176.

Thomas, C. 1870. Descriptions of grasshoppers from Colorado. *Proceedings of the Academy of Natural Sciences of Philadelphia* 1870: 74–84.

—— 1871a. A list and descriptions of new species of Orthoptera. *Preliminary Report, United States Geological Survey of Wyoming & Adjacent Territories* 2, 1870: 265–284.

—— 1871b. Contribution to Orthopterology. *Proceedings of the Academy of Natural Sciences of Philadelphia* 23: 149–153.

—— 1872. Notes on the Saltatorial Orthoptera of the Rocky Mountain Regions. In F. V. Hayden, *Preliminary Report of the U. S. Geological survey of Montana and portions of adjacent territories. 5th Annual Report of Progress, Part IV. Zoology and Botany*, pp. 423–466.

—— 1873a. Notes on Orthoptera. *Sixth Annual Report of the United States Geological Survey of Montana, Idaho, Wyoming and Utah*, pp. 719–725.

—— 1873b. Synopsis of the Acrididae of North America. *Report of the United States Geological Survey of the Territories* 5: 1–258.

—— 1873c. Descriptions of new species of Orthoptera collected in Nevada, Utah, and Arizona, by the expedition under Lieutenant G. M. Wheeler. *Proceedings of the Academy of Natural Sciences of Philadelphia* 25: 295–297.

—— 1874. Descriptions of some new Orthoptera, and notes on some species but little known. *Bulletin of the United States Geological and Geographical Survey of Territories* 1 (2): 63–71.

—— 1875. Report upon the collections of Orthoptera made in portions of

Nevada, Utah, California, Colorado, New Mexico and Arizona during the years 1871–1874. *Wheeler's Report on Geological and Geographical Exploration and Surveys West of the 100th Meridian* 5:
843–908.

——— 1876a. A list of Orthoptera collected by J. Duncan Putnam, of Davenport, Iowa during the summers of 1872–1875, chiefly in Colorado,
Utah, and Wyoming Territories. *Proceedings of the Davenport Academy of Natural Sciences* 1: 249–264.

——— 1876b. A list of the Orthoptera of Illinois. *Illinois Museum of Natural
History Bulletin* no. 1: 59–69.

Thompson, R. M., and G. M. Buxton. 1964. An index of the Acridoidea
(Orthoptera) of California, with selected references. *Bureau of Entomology, California Department of Agriculture, Occasional Papers* 5:
1–62.

Tinkham, E. R. 1948. Faunistic and ecological studies on the Orthoptera of
the Big Bend region of Trans-Pecos, Texas. *American Midland Naturalist* 40: 521–663.

——— 1961. Studies in Nearctic desert sand dune Orthoptera. Part III: A
new species of *Cibolacris* from northern Chihuahua, Mexico. *The Great
Basin Naturalist* 21: 29–33.

Uhler, P. R. 1864. Orthopterological contributions. *Proceedings of the Entomological Society of Philadelphia* 1864: 543–555.

——— 1877. Report upon the insects collected by P. R. Uhler during the
explorations of 1875. *Bulletin of the U. S. Geological and Geographical
Survey of the Territories* 3 (4): 765–802.

Uvarov, B. P. 1925. Notes on the Orthoptera in the British Museum. 4.
Identification of types of Acrididae preserved in the museum. *Transactions of the Entomological Society of London* 1924 (3, 4): 265–301.

——— 1940a. Eleven new generic names in Orthoptera. *Annals and Magazine of Natural History* 6 (11): 377–380.

——— 1940b. Twenty-eight new generic names in Orthoptera. *Annals and
Magazine of Natural History* 5 (11): 173–176.

——— 1940c. Twenty-four new generic names in Orthoptera. *Annals and
Magazine of Natural History* 6 (11): 112–117.

——— 1966. *Grasshoppers and locusts. A handbook of general acridology.*
Cambridge, Eng.

Vickery, V. R. 1964. The validity of the name *curtipennis* (Harris) for North
American *Chorthippus* (Orthoptera: Acrididae). *Canadian Entomologist* 96: 1537–1548.

——— 1967. Distribution and variation in North American *Chorthippus*
(Orthoptera: Acrididae: Gomphocerinae). *Annals of the Entomological
Society of Quebec* 12: 100–132.

——— 1969a. A new genus and species of Orthoptera (Acridoidea: Acrididae) from Sonora, Mexico. *Canadian Entomologist* 101: 1223–1227.

——— 1969b. Two new species of subarctic American Orthoptera. *Entomological News* 80: 265–272.

Walker, E. M. 1909. On the Orthoptera of Northern Ontario. *Canadian Entomologist* 41: 137–144, 173–178, 205–212.

Walker, F. 1870–1871. *Catalogue of the specimens of Dermaptera and Saltatoria in the collection of the British Museum.* Part 3, pp. 425–604; part 4, pp. 606–801; part 5, pp. 810–850; supplement, pp. 1–116.

White, M. J. D., and N. H. Nickerson 1951. Structural heterozygosity in a very rare species of grasshopper. *American Naturalist* 85: 239–246.

Taxonomic Index

Names in bold type are valid species and genera; names in italics are junior synonyms. Following each species and genus is the author. In parentheses after the species name is the sequence of genera to which it has been assigned; the last name is the present assignment.

abdominalis Thomas (*Chrysochraon, Neopodismopsis,* **Chloealtis**), 42
abortivus Bruner (*Mesochloa,* **Eritettix**), 110
Acantherus Scudder and Cockerell, 200
Acentetus McNeill (emendation of *Akentetus* McNeill) = **Amphitornus** McNeill, 225
Achurum Saussure, 185
acridoides Stål (*Truxalis*) = **Achurum sumichrasti**, 231
acris Rehn and Hebard (*Paropomala*) = **Paropomala pallida**, 245
Acrocara Scudder = **Acrolophitus** Thomas, 225
Acrolophitus Thomas, 201
aculeata Rehn (*Orphulella*), 91
acuta Rehn (*Orphulina*) = **Orphulina balloui**, 244
acuticornis Bruner (*Syrbula*) = **Syrbula montezuma**, 249
acutus Morse (*Stenobothrus*) = **Chorthippus curtipennis**, 236
admirabilis Uhler (*Stenobothrus, Syrbula*), 176
aequalis Scudder (*Stenobothrus*) = **Orphulella speciosa**, 244
Aeropedellus Hebard, 53
affinis Morse (*Cordillacris*) = **Cordillacris occipitalis**, 236

affinis Scudder (*Orphulella*) = **Orphulella pelidna**, 242
Ageneotettix McNeill, 136
Akentetus McNeill = **Amphitornus** McNeill, 225
alacris Scudder (*Mermiria*) = **Mermiria picta**, 239
Alpha Bruner, preoccupied by *Alpha* Saussure = **Cordillacris** Rehn, 226
alticola Rehn (*Platybothrus, Bruneria*) = **Stenobothrus shastanus**, 248
altus Scudder (*Oenomus*) = **Chloealtis gracilis**, 235
Amblytropidia Stål, 168
Amphitornus McNeill, 116
apache Rehn and Hebard (*Cordillacris*) = **Cordillacris crenulata**, 237
aphelocoryphus Jago (**Silvitettix**), 77
aptera Scudder (*Opomala*) = **Pseudopomala brachyptera**, 247
arctatus Walker (*Stenobothrus*) = **Orphulella punctata**, 243
arcticus Hebard (**Aeropedellus**), 55
arenosus Hancock (*Ageneotettix*) = **Ageneotettix deorum**, 232
argentatus Bruner (**Bootettix**), 207
aridus Bruner (*Thrincus*) = **Cibolacris parviceps**, 236
arizonensis Bruner (*Papagoa, Mermiria*) = **Mermiria texana**, 240